FREE Test Taking Tips DVD Offer

To help us better serve you, we have developed a Test Taking Tips DVD that we would like to give you for FREE. **This DVD covers world-class test taking tips that you can use to be even more successful when you are taking your test.**

All that we ask is that you email us your feedback about your study guide. Please let us know what you thought about it – whether that is good, bad or indifferent.

To get your **FREE Test Taking Tips DVD**, email freedvd@studyguideteam.com with "FREE DVD" in the subject line and the following information in the body of the email:

    a. The title of your study guide.

    b. Your product rating on a scale of 1-5, with 5 being the highest rating.

    c. Your feedback about the study guide. What did you think of it?

    d. Your full name and shipping address to send your free DVD.

If you have any questions or concerns, please don't hesitate to contact us at freedvd@studyguideteam.com.

Thanks again!

SAT Biology Subject Test 2020 & 2021

SAT Bio E/M Subject Test and Practice Exam Questions [2nd Edition]

Test Prep Books

Interested in buying more than 10 copies of our product? Contact us about bulk discounts:
bulkorders@studyguideteam.com

ISBN 13: 9781628458930
ISBN 10: 1628458933

Table of Contents

Quick Overview

As you draw closer to taking your exam, effective preparation becomes more and more important. Thankfully, you have this study guide to help you get ready. Use this guide to help keep your studying on track and refer to it often.

This study guide contains several key sections that will help you be successful on your exam. The guide contains tips for what you should do the night before and the day of the test. Also included are test-taking tips. Knowing the right information is not always enough. Many well-prepared test takers struggle with exams. These tips will help equip you to accurately read, assess, and answer test questions.

A large part of the guide is devoted to showing you what content to expect on the exam and to helping you better understand that content. In this guide are practice test questions so that you can see how well you have grasped the content. Then, answer explanations are provided so that you can understand why you missed certain questions.

Don't try to cram the night before you take your exam. This is not a wise strategy for a few reasons. First, your retention of the information will be low. Your time would be better used by reviewing information you already know rather than trying to learn a lot of new information. Second, you will likely become stressed as you try to gain a large amount of knowledge in a short amount of time. Third, you will be depriving yourself of sleep. So be sure to go to bed at a reasonable time the night before. Being well-rested helps you focus and remain calm.

Be sure to eat a substantial breakfast the morning of the exam. If you are taking the exam in the afternoon, be sure to have a good lunch as well. Being hungry is distracting and can make it difficult to focus. You have hopefully spent lots of time preparing for the exam. Don't let an empty stomach get in the way of success!

When travelling to the testing center, leave earlier than needed. That way, you have a buffer in case you experience any delays. This will help you remain calm and will keep you from missing your appointment time at the testing center.

Be sure to pace yourself during the exam. Don't try to rush through the exam. There is no need to risk performing poorly on the exam just so you can leave the testing center early. Allow yourself to use all of the allotted time if needed.

Remain positive while taking the exam even if you feel like you are performing poorly. Thinking about the content you should have mastered will not help you perform better on the exam.

Once the exam is complete, take some time to relax. Even if you feel that you need to take the exam again, you will be well served by some down time before you begin studying again. It's often easier to convince yourself to study if you know that it will come with a reward!

Test-Taking Strategies

1. Predicting the Answer

When you feel confident in your preparation for a multiple-choice test, try predicting the answer before reading the answer choices. This is especially useful on questions that test objective factual knowledge. By predicting the answer before reading the available choices, you eliminate the possibility that you will be distracted or led astray by an incorrect answer choice. You will feel more confident in your selection if you read the question, predict the answer, and then find your prediction among the answer choices. After using this strategy, be sure to still read all of the answer choices carefully and completely. If you feel unprepared, you should not attempt to predict the answers. This would be a waste of time and an opportunity for your mind to wander in the wrong direction.

2. Reading the Whole Question

Too often, test takers scan a multiple-choice question, recognize a few familiar words, and immediately jump to the answer choices. Test authors are aware of this common impatience, and they will sometimes prey upon it. For instance, a test author might subtly turn the question into a negative, or he or she might redirect the focus of the question right at the end. The only way to avoid falling into these traps is to read the entirety of the question carefully before reading the answer choices.

3. Looking for Wrong Answers

Long and complicated multiple-choice questions can be intimidating. One way to simplify a difficult multiple-choice question is to eliminate all of the answer choices that are clearly wrong. In most sets of answers, there will be at least one selection that can be dismissed right away. If the test is administered on paper, the test taker could draw a line through it to indicate that it may be ignored; otherwise, the test taker will have to perform this operation mentally or on scratch paper. In either case, once the obviously incorrect answers have been eliminated, the remaining choices may be considered. Sometimes identifying the clearly wrong answers will give the test taker some information about the correct answer. For instance, if one of the remaining answer choices is a direct opposite of one of the eliminated answer choices, it may well be the correct answer. The opposite of obviously wrong is obviously right! Of course, this is not always the case. Some answers are obviously incorrect simply because they are irrelevant to the question being asked. Still, identifying and eliminating some incorrect answer choices is a good way to simplify a multiple-choice question.

4. Don't Overanalyze

Anxious test takers often overanalyze questions. When you are nervous, your brain will often run wild, causing you to make associations and discover clues that don't actually exist. If you feel that this may be a problem for you, do whatever you can to slow down during the test. Try taking a deep breath or counting to ten. As you read and consider the question, restrict yourself to the particular words used by the author. Avoid thought tangents about what the author *really* meant, or what he or she was *trying* to say. The only things that matter on a multiple-choice test are the words that are actually in the question. You must avoid reading too much into a multiple-choice question, or supposing that the writer meant something other than what he or she wrote.

5. No Need for Panic

It is wise to learn as many strategies as possible before taking a multiple-choice test, but it is likely that you will come across a few questions for which you simply don't know the answer. In this situation, avoid panicking. Because most multiple-choice tests include dozens of questions, the relative value of a single wrong answer is small. As much as possible, you should compartmentalize each question on a multiple-choice test. In other words, you should not allow your feelings about one question to affect your success on the others. When you find a question that you either don't understand or don't know how to answer, just take a deep breath and do your best. Read the entire question slowly and carefully. Try rephrasing the question a couple of different ways. Then, read all of the answer choices carefully. After eliminating obviously wrong answers, make a selection and move on to the next question.

6. Confusing Answer Choices

When working on a difficult multiple-choice question, there may be a tendency to focus on the answer choices that are the easiest to understand. Many people, whether consciously or not, gravitate to the answer choices that require the least concentration, knowledge, and memory. This is a mistake. When you come across an answer choice that is confusing, you should give it extra attention. A question might be confusing because you do not know the subject matter to which it refers. If this is the case, don't eliminate the answer before you have affirmatively settled on another. When you come across an answer choice of this type, set it aside as you look at the remaining choices. If you can confidently assert that one of the other choices is correct, you can leave the confusing answer aside. Otherwise, you will need to take a moment to try to better understand the confusing answer choice. Rephrasing is one way to tease out the sense of a confusing answer choice.

7. Your First Instinct

Many people struggle with multiple-choice tests because they overthink the questions. If you have studied sufficiently for the test, you should be prepared to trust your first instinct once you have carefully and completely read the question and all of the answer choices. There is a great deal of research suggesting that the mind can come to the correct conclusion very quickly once it has obtained all of the relevant information. At times, it may seem to you as if your intuition is working faster even than your reasoning mind. This may in fact be true. The knowledge you obtain while studying may be retrieved from your subconscious before you have a chance to work out the associations that support it. Verify your instinct by working out the reasons that it should be trusted.

8. Key Words

Many test takers struggle with multiple-choice questions because they have poor reading comprehension skills. Quickly reading and understanding a multiple-choice question requires a mixture of skill and experience. To help with this, try jotting down a few key words and phrases on a piece of scrap paper. Doing this concentrates the process of reading and forces the mind to weigh the relative importance of the question's parts. In selecting words and phrases to write down, the test taker thinks about the question more deeply and carefully. This is especially true for multiple-choice questions that are preceded by a long prompt.

9. Subtle Negatives

One of the oldest tricks in the multiple-choice test writer's book is to subtly reverse the meaning of a question with a word like *not* or *except*. If you are not paying attention to each word in the question, you can easily be led astray by this trick. For instance, a common question format is, "Which of the following is…?" Obviously, if the question instead is, "Which of the following is not…?," then the answer will be quite different. Even worse, the test makers are aware of the potential for this mistake and will include one answer choice that would be correct if the question were not negated or reversed. A test taker who misses the reversal will find what he or she believes to be a correct answer and will be so confident that he or she will fail to reread the question and discover the original error. The only way to avoid this is to practice a wide variety of multiple-choice questions and to pay close attention to each and every word.

10. Reading Every Answer Choice

It may seem obvious, but you should always read every one of the answer choices! Too many test takers fall into the habit of scanning the question and assuming that they understand the question because they recognize a few key words. From there, they pick the first answer choice that answers the question they believe they have read. Test takers who read all of the answer choices might discover that one of the latter answer choices is actually *more* correct. Moreover, reading all of the answer choices can remind you of facts related to the question that can help you arrive at the correct answer. Sometimes, a misstatement or incorrect detail in one of the latter answer choices will trigger your memory of the subject and will enable you to find the right answer. Failing to read all of the answer choices is like not reading all of the items on a restaurant menu: you might miss out on the perfect choice.

11. Spot the Hedges

One of the keys to success on multiple-choice tests is paying close attention to every word. This is never truer than with words like almost, most, some, and sometimes. These words are called "hedges" because they indicate that a statement is not totally true or not true in every place and time. An absolute statement will contain no hedges, but in many subjects, the answers are not always straightforward or absolute. There are always exceptions to the rules in these subjects. For this reason, you should favor those multiple-choice questions that contain hedging language. The presence of qualifying words indicates that the author is taking special care with his or her words, which is certainly important when composing the right answer. After all, there are many ways to be wrong, but there is only one way to be right! For this reason, it is wise to avoid answers that are absolute when taking a multiple-choice test. An absolute answer is one that says things are either all one way or all another. They often include words like *every*, *always*, *best*, and *never*. If you are taking a multiple-choice test in a subject that doesn't lend itself to absolute answers, be on your guard if you see any of these words.

12. Long Answers

In many subject areas, the answers are not simple. As already mentioned, the right answer often requires hedges. Another common feature of the answers to a complex or subjective question are qualifying clauses, which are groups of words that subtly modify the meaning of the sentence. If the question or answer choice describes a rule to which there are exceptions or the subject matter is complicated, ambiguous, or confusing, the correct answer will require many words in order to be expressed clearly and accurately. In essence, you should not be deterred by answer choices that seem excessively long. Oftentimes, the author of the text will not be able to write the correct answer without

offering some qualifications and modifications. Your job is to read the answer choices thoroughly and completely and to select the one that most accurately and precisely answers the question.

13. Restating to Understand

Sometimes, a question on a multiple-choice test is difficult not because of what it asks but because of how it is written. If this is the case, restate the question or answer choice in different words. This process serves a couple of important purposes. First, it forces you to concentrate on the core of the question. In order to rephrase the question accurately, you have to understand it well. Rephrasing the question will concentrate your mind on the key words and ideas. Second, it will present the information to your mind in a fresh way. This process may trigger your memory and render some useful scrap of information picked up while studying.

14. True Statements

Sometimes an answer choice will be true in itself, but it does not answer the question. This is one of the main reasons why it is essential to read the question carefully and completely before proceeding to the answer choices. Too often, test takers skip ahead to the answer choices and look for true statements. Having found one of these, they are content to select it without reference to the question above. Obviously, this provides an easy way for test makers to play tricks. The savvy test taker will always read the entire question before turning to the answer choices. Then, having settled on a correct answer choice, he or she will refer to the original question and ensure that the selected answer is relevant. The mistake of choosing a correct-but-irrelevant answer choice is especially common on questions related to specific pieces of objective knowledge. A prepared test taker will have a wealth of factual knowledge at his or her disposal, and should not be careless in its application.

15. No Patterns

One of the more dangerous ideas that circulates about multiple-choice tests is that the correct answers tend to fall into patterns. These erroneous ideas range from a belief that B and C are the most common right answers, to the idea that an unprepared test-taker should answer "A-B-A-C-A-D-A-B-A." It cannot be emphasized enough that pattern-seeking of this type is exactly the WRONG way to approach a multiple-choice test. To begin with, it is highly unlikely that the test maker will plot the correct answers according to some predetermined pattern. The questions are scrambled and delivered in a random order. Furthermore, even if the test maker was following a pattern in the assignation of correct answers, there is no reason why the test taker would know which pattern he or she was using. Any attempt to discern a pattern in the answer choices is a waste of time and a distraction from the real work of taking the test. A test taker would be much better served by extra preparation before the test than by reliance on a pattern in the answers.

FREE DVD OFFER

Don't forget that doing well on your exam includes both understanding the test content and understanding how to use what you know to do well on the test. We offer a completely FREE Test Taking Tips DVD that covers world class test taking tips that you can use to be even more successful when you are taking your test.

All that we ask is that you email us your feedback about your study guide. To get your **FREE Test Taking Tips DVD**, email freedvd@studyguideteam.com with "FREE DVD" in the subject line and the following information in the body of the email:

- The title of your study guide.
- Your product rating on a scale of 1-5, with 5 being the highest rating.
- Your feedback about the study guide. What did you think of it?
- Your full name and shipping address to send your free DVD.

Introduction

Function of the Test

The two sections offered on the SAT Subject Test in Biology are Biology Ecological or Biology Molecular. High school students take the exam to showcase their interest in science to the colleges they are applying to. Some colleges use the SAT Subject Test in Biology to place students in appropriate courses in college, allowing them to pass over the basics into more complex courses. Note that in New York State, some people may use SAT Subject Test scores as a substitute for a Regents examination score. For both ecological and molecular, it's recommended that test takers have at least a year of an introductory course in biology, a year of a course in algebra, and experience in a laboratory setting.

Test Administration

To find out where the SAT Subject Test in Biology is given, go to the website at collegereadiness.collegeboard.org. There you can find exam dates for the current year as well as up to three years in advance. The dates are listed by subject and give a deadline to register for each date. Currently, there are six dates listed each year for taking the subject tests, all of which offer the Biology E/M subject tests. The dates for the exams are offered in August, October, November, December, May, and June. Note that you cannot take the SAT and an SAT Subject Test on the same date. The College Board website also has a test date finder, where you enter in the test date and your country, state, and city. The results bring up testing centers in your area or the area(s) nearest you.

Test Format

The SAT Biology Subject test is sixty minutes long and has eighty multiple choice questions. Sixty of the questions are traditional Biology E/M questions, and then the rest of the twenty questions are specialized. The skills anticipated for both the Biology E/M exams are fundamental concepts and knowledge (30% of the test), application (35% of the test), and interpretation (35% of the test). These skills include understanding major principles in biology and applying principles to solve problems in biology, comprehending algebraic concepts and applying those concepts to word problems, drawing observations from experiments and working with data, and knowledge of the metric system of units.

The following table shows how the content is broken up for both Ecology and Molecular:

Content	Percent of Test	
	Ecological	Molecular
Cellular and Molecular Biology	15%	27%
Ecology	23%	13%
Genetics	15%	20%
Organismal Biology	25%	25%
Evolution and Diversity	22%	15%

Note that if you want to take both Biology E and M, you have to take them on separate testing dates. When you sit down to take the exam, you must fill in either "Biology E" or "Biology M" to indicate which test you want to take. Biology E is for test takers who are more knowledgeable of biological

communities, energy flow, and populations. Biology M is for test takers who are more knowledgeable in biochemistry and cellular structure and processes.

Scoring

In 2016, 72,196 students took the SAT Subject Test in Biology E/M. 31,965 took Ecological and 40,231 took Molecular. Out of a 200 to 800-point scale, the mean score was 616 for Ecological and 647 for Molecular. The standard deviation on the Ecological test was 114, and the standard deviation on the Molecular test was 110. This calculates the spread of grades around the average score. A "good score" for the subject test in biology depends on the college you are applying to. Although many colleges are flexible and look at SAT scores along with other factors for admissions, highly selective schools prefer to see a score in the 700s or above. Note that you do not have to take SAT Subject Tests for admission to most colleges; even many highly selective schools recommend but do not require the exam.

Recent Developments

The only new developments for the subject tests are changes to the testing calendar. The subject tests are offered on certain Saturdays in August, October, November, December, May, and June. For students who have to miss the Saturday test for religious reasons, the test is also offered on Sundays. Exam dates are no longer offered in January. Testing is available internationally, and subject tests will be offered outside of the U.S. and U.S. territories in October, November, December, May, and June.

Cellular and Molecular Biology

Cell Structure and Organization

Structure and Function of Cell Membranes

All cells are surrounded by a cell membrane that is formed from two layers of phospholipids. **Phospholipids** are two fatty acid chains connected to a glycerol molecule with a phosphate group. The membrane is amphiphilic because the fatty acid chains are hydrophobic and the phosphate group is hydrophilic. This creates a unique environment that protects the cell's inner contents while still allowing material to pass through the membrane. Because the outside of a cell, known as the **extracellular space**, and the inside of a cell, the **intercellular space**, are aqueous, the lipid bilayer forms with the two phospholipid heads facing the outside and the inside of the cell. This allows the phospholipids to interact with water; the fatty acid tails face the middle, so they can interact with each other and avoid water.

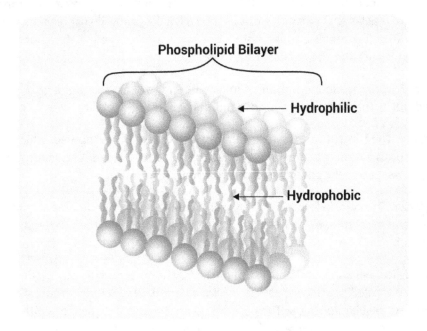

Molecules can pass through the cell membrane by either active or passive transport. **Active transport** requires chemical energy to move molecules in or out of the cell from areas of low concentration to areas of high concentration, or in instances where the molecules would not be able to pass through by themselves, such as with large non-lipid soluble molecules. Ions, amino acids, and complex sugars use active transport mechanisms. Molecules that are soluble in lipids, water, and oxygen use **passive transport** to move in and out of the cell, which means that cellular energy is not required for their movement. Examples of passive transport include diffusion, facilitated diffusion, and osmosis. **Diffusion** is the net movement of particles from an area of high concentration to lower concentration. **Facilitated diffusion** is the movement of molecules through cell membranes with the use of special transport proteins. Finally, **osmosis** is the movement of water molecules across partially permeable membranes.

Structure and Function of Animal and Plant Cell Organelles

Animal and plant cells contain many of the same or similar **organelles**, which are membrane enclosed structures that each have a specific function; however, there are a few organelles that are unique to either one or the other general cell type. The following cell organelles are found in both animal and plant cells, unless otherwise noted in their description:

- *Nucleus*: The nucleus consists of three parts: the nuclear envelope, the nucleolus, and chromatin. The **nuclear envelope** is the double membrane that surrounds the nucleus and separates its contents from the rest of the cell. The **nucleolus** produces ribosomes. **Chromatin** consists of DNA and protein, which form chromosomes that contain genetic information. Most cells have only one nucleus; however, some cells, such as skeletal muscle cells, have multiple nuclei.

- *Endoplasmic reticulum (ER)*: The ER is a network of membranous sacs and tubes that is responsible for membrane synthesis. It is also responsible for packaging and transporting proteins into vesicles that can move out of the cell. It folds and transports other proteins to the Golgi apparatus. It contains both smooth and rough regions; the rough regions have ribosomes attached, which are the sites of protein synthesis.

- *Flagellum*: Flagella are found only in animal cells. They are made up of a cluster of microtubules projected out of the plasma membrane, and they aid in cell mobility.

- *Centrosome*: The centrosome is the area of the cell where **microtubules**, which are filaments that are responsible for movement in the cell, begin to be formed. Each centrosome contains two centrioles. Each cell contains one centrosome.

- *Cytoskeleton*: The cytoskeleton in animal cells is made up of microfilaments, intermediate filaments, and microtubules. In plant cells, the cytoskeleton is made up of only microfilaments and microtubules. These structures reinforce the cell's shape and aid in cell movement.

- *Microvilli*: Microvilli are found only in animal cells. They are protrusions in the cell membrane that increase the cell's surface area. They have a variety of functions, including absorption, secretion, and cellular adhesion. They are found on the apical surface of epithelial cells, such as in the small intestine. They are also located on the plasma surface of a female's eggs to help anchor sperm that are attempting fertilization.

- *Peroxisome*: A peroxisome contains enzymes that are involved in many of the cell's metabolic functions, one of the most important being the breakdown of very long chain fatty acids. Peroxisomes produces hydrogen peroxide as a byproduct of these processes and then converts the hydrogen peroxide to water. There are many peroxisomes in each cell.

- *Mitochondrion*: The mitochondrion is often called the powerhouse of the cell and is one of the most important structures for maintaining regular cell function. It is where aerobic cellular respiration occurs and where most of the cell's **adenosine triphosphate (ATP)** is generated. The number of mitochondria in a cell varies greatly from organism to organism, and from cell to cell. In human cells, the number of mitochondria can vary from zero in a red blood cell, to 2000 in a liver cell.

- *Lysosome*: Lysosomes are responsible for digestion and can hydrolyze macromolecules. There are many lysosomes in each cell.

- *Golgi apparatus*: The Golgi apparatus is responsible for the composition, modification, organization, and secretion of cell products. Because of its large size, it was actually one of the first organelles to be studied in detail. There are many Golgi apparatuses in each cell.

- *Ribosomes*: Ribosomes are found either free in the cytosol, bound to the rough ER, or bound to the nuclear envelope. They manufacture proteins within the cell.

- *Plasmodesmata*: The plasmodesmata are found only in plant cells. They are cytoplasmic channels, or tunnels, that go through the cell wall and connect the cytoplasm of adjacent cells.

- *Chloroplast*: Chloroplasts are found only in plant cells. They are responsible for **photosynthesis**, which is the process of converting sunlight to chemical energy that can be stored and used later to drive cellular activities.

- *Central vacuole*: A central vacuole is found only in plant cells. It is responsible for storing material and waste. This is the only vacuole found in a plant cell.

- *Plasma membrane*: The plasma membrane is a phospholipid bilayer that encloses the cell.

- *Cell wall*: Cell walls are only present in plant cells. The cell wall is made up of strong fibrous substances including cellulose and other polysaccharides, and protein. It is a layer outside of the plasma membrane, which protects the cell from mechanical damage and helps maintain the cell's shape.

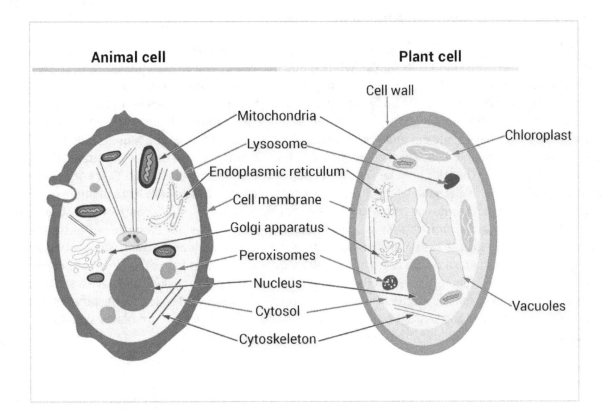

Levels of Organization

There are about two hundred different types of cells in the human body. Cells group together to form *biological tissues*, and tissues combine to form *organs*, such as the heart and kidneys. Organs that work together to perform vital functions of the human body form **organ systems**. There are eleven organ systems in the human body: skeletal, muscular, urinary, nervous, digestive, endocrine, reproductive, respiratory, cardiovascular, integumentary, and lymphatic. Although each system has its own unique function, they all rely on each other, either directly or indirectly, to operate properly.

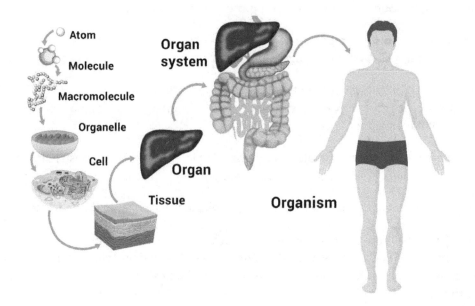

Major Features of Common Animal Cell Types

The most common animal cell types are blood, muscle, nerve, epithelial, and gamete cells. The three main blood cells are **red blood cells (RBCs)**, **white blood cells (WBCs)**, and **platelets**. RBCs transport oxygen and carbon dioxide through the body. They do not have a nucleus and they live for about 120 days in the blood. WBCs defend the body against diseases. They do have a nucleus and live for only three to four days in the human body. Platelets help with the formation of blood clots following an injury. They do not have a nucleus and live for about eight days after formation. **Muscle cells** are long, tubular cells that form muscles, which are responsible for movement in the body. On average, they live for about fifteen years, but this number is highly dependent on the individual body. There are three main types of muscle tissue: skeletal, cardiac, and smooth. **Skeletal muscle cells** have multiple nuclei and are the only voluntary muscle cell, which means that the brain consciously controls the movement of skeletal muscle. **Cardiac muscle cells** are only found in the heart; they have a single nucleus and are involuntary. **Smooth muscle cells** make up the walls of the blood vessels and organs. They have a single nucleus and are involuntary. **Nerve cells** conduct electrical impulses that help send information and instructions from the brain to the rest of the body. They contain a single nucleus and have a specialized membrane that allows for this electrical signaling between cells. **Epithelial cells** cover exposed surfaces, and line internal cavities and passageways. **Gametes** are specialized cells that are responsible for reproduction. In the human body, the gametes are the egg and the sperm.

—

Prokaryotes and Eukaryotes

There are two distinct types of cells that make up most living organisms: **prokaryotic** and **eukaryotic**. Both types of cells are enclosed by a cell membrane, which is selectively permeable. Selective permeability means essentially that it is a gatekeeper, allowing certain molecules and ions in and out, and keeping unwanted ones at bay, at least until they are ready for use. They both contain ribosomes and DNA. One major difference between these types of cells is that in eukaryotic cells, the cell's DNA is enclosed in a membrane-bound nucleus, whereas in prokaryotic cells, the cell's DNA is in a region—called the **nucleoid**—that is not enclosed by a membrane. Another major difference is that eukaryotic cells contain organelles, while prokaryotic cells do not have membrane-bound organelles.

Prokaryotic cells include **bacteria** and **archaea**. They do not have a nucleus or any membrane-bound organelles, are unicellular organisms, and are generally very small in size. Eukaryotic cells include animal, plant, fungus, and protist cells. **Fungi** are unicellular microorganisms such as yeasts, molds, and mushrooms. Their distinguishing characteristic is the chitin that is in their cell walls. **Protists** are organisms that are not classified as animals, plants, or fungi; they are unicellular; and they do not form tissues.

Mitosis

Cell Cycle

The Cell Cycle

Mitotic Phase — Cytokinesis, Telophase, Anaphase, Metaphase, Prophase (Mitosis)

Interphase — G_1: Centrioles replicate; S: DNA replication; G_2: Final growth and activity before mitosis

The **cell cycle** is the process by which a cell divides and duplicates itself. There are two processes by which a cell can divide itself: mitosis and meiosis. In **mitosis**, the daughter cells that are produced from

parental cell division are identical to each other and the parent. **Meiosis** is a unique process that involves two stages of cell division and produces **haploid cells**, which are cells containing only one set of chromosomes, from **diploid parent cells**, which are cells containing two sets of chromosomes.

Interphase

The majority of the time spent in the cell cycle is a step called **interphase**. The cell undergoes regular metabolic activities and growth in this time, increasing the number of proteins and organelles in the cell. During interphase, the cell prepares itself for the mitotic phase in three phases called the Gap 1 (G_1), Synthesis (S), and Gap 2 (G_2) phases. During the G_1 phase, the centrosome is replicated and the cell grows to nearly double its original size. This is often the longest part of interphase and has the largest differences in time between cell types. The S phase is when the DNA in the cell is replicated, creating a copy of the chromosomes in preparation for being split during nuclear division in the mitotic phase. The last part of interphase is the G_2 phase, where the cell continues to grow. By the end of interphase, the cell has doubled its genetic material, organelles, and is prepared to divide in the mitotic phase.

Mitosis

Mitosis can be broken down into five stages: prophase, prometaphase, metaphase, anaphase, and telophase.

- *Prophase*: During this phase, the mitotic spindles begin to form from centrosomes and microtubules. As the microtubules lengthen, the centrosomes move farther away from each other. The nucleolus disappears and the chromatin fibers begin to coil up and form chromosomes. Two sister chromatids, which are two copies of one chromosome, are joined together.

- *Prometaphase*: The nuclear envelope begins to break down and the microtubules enter the nuclear area. Each pair of chromatin fibers develops a **kinetochore**, which is a specialized protein structure in the middle of the adjoined fibers. The chromosomes are further condensed.

- *Metaphase*: In this stage, the microtubules are stretched across the cell and the centrosomes are at opposite ends of the cell. The chromosomes align at the **metaphase plate**, which is a plane that is exactly between the two centrosomes. The kinetochore of each chromosome is attached to the kinetochore of the microtubules that are stretching from each centrosome to the metaphase plate.

- *Anaphase*: The sister chromatids break apart, forming full-fledged chromosomes. The two daughter chromosomes move to opposite ends of the cell. The microtubules shorten toward opposite ends of the cell as well, and the cell elongates.

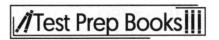

- *Telophase*: Two nuclei form at each end of the cell and nuclear envelopes begin to form around each nucleus. The nucleoli reappear and the chromosomes become less condensed. The microtubules are broken down by the cell and mitosis is complete.

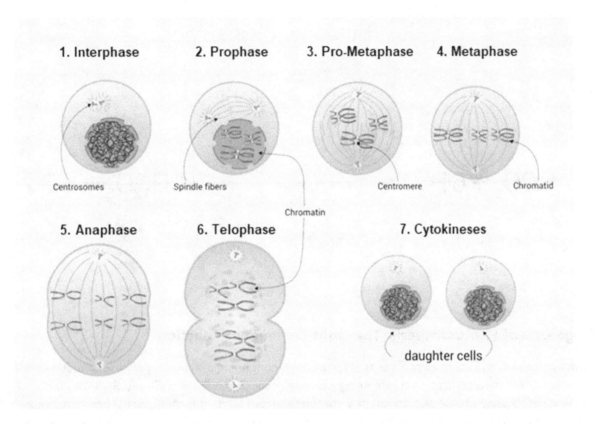

Mitosis has several checkpoints. If a cell exits the cell cycle, it enters a phase called G_0 that consists of non-dividing cells like neurons. G_1 checkpoints prepare and commit cells to entering the cell cycle. S phase proofreads and corrects mistakes. There is also a G_2 checkpoint: as a cell progresses through G_1, S, and G_2, the cyclin protein accumulates. When it becomes abundant, it binds with a **cyclin dependent kinase (CDK)** to form **Maturation Promoting Factor (MPF)**, an activating protein complex that facilitates mitosis. As mitosis is completed, the cyclin is degraded, the MPF complex disassembles, and the cell cycle begins once again.

Cytokinesis

Cytokinesis is the division of cytoplasm that occurs immediately following the division of genetic material during cellular reproduction. The process of mitosis or meiosis, followed by cytokinesis, finishes up the cell cycle.

Photosynthesis

Photosynthesis is the process of converting light energy into chemical energy, which is then stored in sugar and other organic molecules. It can be divided into two stages called the **light reactions** and the **Calvin cycle**. The photosynthetic process takes place in the chloroplasts in plants. Inside the chloroplast, there are membranous sacs called **thylakoids**. **Chlorophyll** is a green pigment that lives in the thylakoid membranes, absorbs photons from light, and starts an **electron transport chain (ETC)** in order to produce energy in the form of ATP and NADPH. The ATP and NADPH produced from the light reactions are used as energy to form organic molecules in the Calvin cycle.

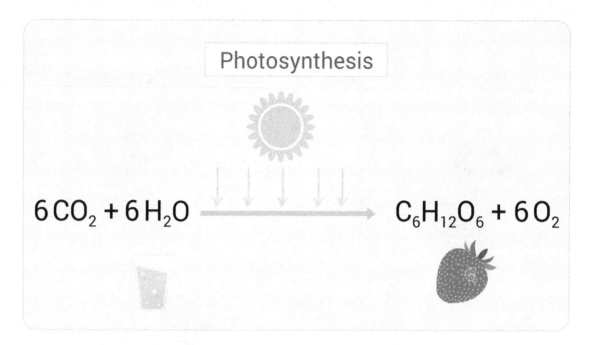

Stage One of Photosynthesis: The Light-Dependent Reaction

Photosynthesis is the complex process that chlorophyll-containing organisms perform to make their own food, which will then be used to create energy through cellular respiration. Chlorophyll is a green pigment responsible for the absorption of a photon (a unit of light), which provides the energy required to begin photosynthesis. Chlorophyll is present in the eukaryotic cells of plants and plant-like protists such as green algae.

Photosynthesis begins with stage one, also known as the light dependent reaction, which results in the creation of energy in the form of ATP and NADPH. It is light dependent because it requires the energy provided by a photon. It occurs in a complicated series of steps, which are outlined here:

1. Light from the sun (in the form of a photon) strikes a molecule of chlorophyll that is embedded within and around the photosystems lodged within the thylakoid membrane of the chloroplast.

2. The photon excites an electron located in **Photosystem II (PSII)**, which is the first of four protein complexes within the membrane. PSII absorbs photons with a wavelength of 680 nanometers in protein 680 (p680). This excited electron jumps into the primary electron acceptor in the center of PSII, where it is picked up by an electron carrier. The electron serves to transfer energy.

3. Meanwhile, PSII takes a molecule of water and splits it into hydrogen and oxygen, stealing an electron from hydrogen to replace the one it just lost, and releasing oxygen and protons (H^+). This process is called **hydrolysis**.

4. The excited electron travels to the next protein complex, the cytochrome complex — the intermediary between PSII and **Photosystem I (PSI)**—which uses the energy from the electron to

pump a proton across the thylakoid membrane into the thylakoid space, creating a positive concentration gradient.

5. The electron, having exhausted all of its energy as it moved along the electron transport chain, splitting water and pumping hydrogen ions, travels to PSI and is re-energized by another photon at a wavelength of 700 nanometers in protein 700 (p700).

6. The re-excited electron is picked up by another electron carrier and taken to the NADP⁺ reductase, an enzyme that uses the energy from the electron to create NADPH, a vessel of stored energy, by accepting hydrogen and two electrons (from the ETC) and donating them to NADP⁺.

7. Meanwhile, the hydrogen protons that build up within the thylakoid space are propelled by their natural inclination to move away from each other (via the repulsion of their charges) and push their way through the ATP synthase. This uses the energy of the proton gradient to add an inorganic phosphate (P_i) to ADP to create ATP.

8. The ATP generated from the electron transport chain and the electrons carried by NADPH are soon invested in the Calvin cycle to create a high-energy glucose.

The overall net reaction of all of the reactions of oxygenic photosynthesis can be seen in the following formula:

$$2\,H_2O + 2\,NADP^+ + 3\,ADP + 3\,P_i \rightarrow O_2 + 2\,NADPH + 3\,ATP$$

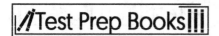

Stage Two of Photosynthesis (The Calvin Cycle): Light-Independent Reactions

The Calvin Cycle

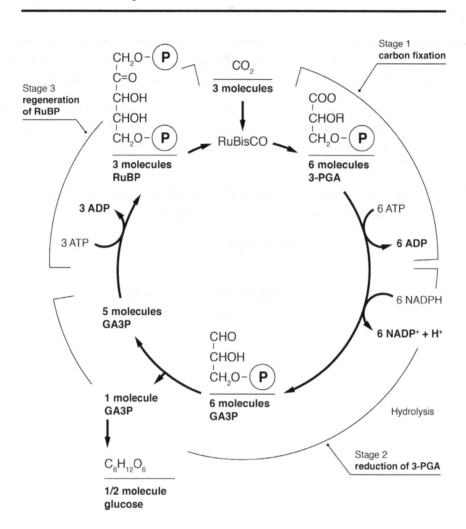

The Calvin cycle is the part of photosynthesis that actually creates glucose. It uses the byproducts ATP and NADPH and the solar energy harnessed in stage one in a series of events described as follows.

1. Carbon dioxide from the atmosphere enters the plant through stomata on the bottoms of its leaves. It then diffuses into the stroma of the chloroplast, located outside the thylakoid membrane.

2. In the exergonic reaction called carbon fixation, the CO_2 then combines with RuBP, a 5-carbon molecule with two phosphate groups, catalyzed by the enzyme called RuBisCO, the most abundant protein in the world. The addition of a sixth carbon causes the molecule to become unstable, so each molecule of RuBP immediately splits into two 3-carbon molecules with a phosphate group

called 3-phosphoglycerate (3-PGA). This process happens three times, resulting in 6 molecules of 3-PGA. (This step is also called the C_3 pathway).

1 turn of the Calvin cycle: $CO_2 + RuBP \rightarrow$ 2 3-PGA

3 turns of the Calvin cycle: $3\ CO_2 + 3\ RuBP \rightarrow$ **6 3-PGA**

3. In an endergonic reaction called reduction, NADPH uses energy from ATP to add a hydrogen to each molecule of 3-phosphoglycerate, turning the six molecules of 3-PGA into a 3-carbon sugar called glyceraldehyde 3-phosphate (G3P). ATP supplies energy by donating a phosphate group (P_i) and becoming ADP, while NADPH loses a hydrogen to become $NADP^+$.

$$6\ ATP + 6\ NADPH + 6\ \text{3-GPA} \rightarrow 6\ G3P + 6\ ADP + 6\ P_i + 6\ NADP^+ + 6\ H^+$$

4. Of the six molecules of G3P created, only one is reserved to make sugar, while the other five molecules are reused in an endergonic reaction called regeneration to replace the three used RuBP molecules. It takes two molecules of G3P to make glucose, and since three turns of the Calvin cycle produce only one G3P, this means it takes six turns of the Calvin cycle to make one 6-carbon molecule of glucose. In the following formulas, remember that most of the G3P is used to renew RuBP.

$$\text{3 turns} = 3\ CO_2 + 3\ RuBP \rightarrow 6\ \text{3-PGA} \rightarrow 6\ G3P \rightarrow 1\ G3P\ \text{exits} \rightarrow 1/2\ C_6H_{12}O_6\ \text{(glucose)}$$

$$\text{6 turns} = 6\ CO_2 + 6\ RuBP \rightarrow 12\ \text{3-PGA} \rightarrow 12\ G3P \rightarrow 2\ G3P\ \text{exit} \rightarrow 1\ C_6H_{12}O_6\ \text{(glucose)}$$

Photorespiration: C3, C4, and CAM Plants

In certain conditions, plants will halt the production of glucose in a process called photorespiration. C_3 plants, the most common, are the most efficient in cool, moist climates. In hot or dry conditions, stomata, the miniscule holes in leaves that enable the transfer of liquid and gases between the plant and its environment, close to conserve water. This can be problematic because it reduces the influx of carbon dioxide. Carbon dioxide becomes scarce, and the oxygen byproduct of the hydrolysis during the light-dependent reaction in PSII builds up, unable to escape. When temperatures increase, RuBisCO has a higher affinity for oxygen and that, combined with the higher O_2 to CO_2 ratio caused by the closed stomata, causes it to bind to O_2 instead of CO_2. This means that carbon cannot be fixed to become glucose, but still it uses ATP to burn up energy, essentially undoing the work of the Calvin cycle. Therefore, photorespiration is resource- and energy-draining, and scientists are not entirely sure of its evolutionary significance. Two alternative systems exist in some plants to avoid photorespiration.

C_4 plants and CAM plants are found in tropical and desert climates. They have mechanisms to avoid photorespiration that they will employ if resources are low. They both take an alternative route to the Calvin cycle, so they actually have the same sugar-producing endgame as C_3 plants. These pathways that circumvent photorespiration either require extra energy (as with C_4 plants) or are not as efficient (as in CAM plants) as functioning C_3 plants, but the fact that they still produce sugar for life-sustaining energy makes them preferable to photorespiration.

C_4 (4-carbon) plants evade photorespiration by using a much more efficient enzyme called PEP carboxylase, which has an affinity for carbon dioxide only. RuBisCO is the preferred enzyme for carbon fixation, since PEP carboxylase requires energy. However, if RuBisCO is being blocked by oxygen in low carbon dioxide situations, PEP carboxylase will bind any circulating carbon dioxide and incorporate it into a C_4 product. This 4-carbon product (called oxaloacetate, or OAA for short) releases carbon dioxide to the Calvin cycle to be used by RuBisCO in carbon fixation. ATP is invested to convert it to a 3-carbon sugar that can bind to PEP carboxylase and repeat the cycle. Basically, PEP carboxylase is acting as a carbon dioxide pump to keep levels high, enabling it to make the high energy glucose.

CAM plants, typically found in deserts, conserve water by keeping their stomata closed during the hot part of the day; this prevents dehydration through transpiration. The stomata open at night to capture and store carbon dioxide in organic compounds. Like C_4 plants, the carbon is fixed into carbon intermediates. During daylight, these compounds are broken up so that released carbon dioxide can bind to RuBisCO and stimulate sugar production via the Calvin cycle. This pathway is beneficial for plants such as cacti that need to conserve water. It is not as efficient due to uncoupling carbon fixation with the rest of the cycle, but for extremely hot environments, it is adaptive.

Cellular Respiration

Cellular respiration is a set of metabolic processes that converts energy from nutrients into ATP, which is the molecule of useable energy for the cell. Respiration can either occur aerobically, using oxygen, or anaerobically, without oxygen. While prokaryotic cells carry out respiration in the cytosol, most of the aerobic respiration in eukaryotic cells occurs in the mitochondria. Glycolysis and ATP-PC (the phosphocreatine system) take place in the cytosol.

Chloroplasts and mitochondria are the organelles responsible for all energy conversion in eukaryotic cells. All eukaryotes have mitochondria, but only plants and green algae have chloroplasts as well.

Glycolysis

The first step of breaking down glucose to make energy is called **glycolysis** (literally "glucose-splitter"), and it occurs in the cytosol of cells.

$$C_6H_{12}O_6 + 2\,ATP \rightarrow 2\,C_3H_4O_3 + 4\,ATP + 2\,NADH$$

$$glucose + activation\ energy \rightarrow 2\ pyruvate + energy + 2\ electron\ carriers$$

As shown in the previous formula, the overall goal of glycolysis is to break glucose in half and into 2 pyruvate molecules. In doing so, it peels off high-energy electrons that were contained in glucose. Two pairs of electrons (stored in phosphate groups) and two hydrogen atoms are invested into the electron carrier NAD+ that behaves just like the electron carrier in photosynthesis by shuttling electrons from one process to the next.

Glycolysis requires 2 ATP of energy investment to proceed to completion, and it produces 4 ATP via substrate level phosphorylation. This net gain of 2 ATP is a small percentage of the total energy produced in aerobic respiration.

In the absence of oxygen, the 2 ATP produced in glycolysis is the only energy gain there is, and fermentation, or anaerobic respiration, will initiate to recycle the electron carrier NAD$^+$. The two chief types of anaerobic respiration/fermentation are lactic acid fermentation and alcohol fermentation. When muscle cells have exceeded their aerobic capacity, they go into anaerobic respiration, which produces lactic acid and 2 net ATP. Yeast undergoes alcohol fermentation, producing carbon dioxide, ethyl alcohol, and 2 net ATP.

Glycolysis is performed via the following steps:

1. Glucose is converted into Glucose-6-Phosphate (G6P) via the enzyme hexokinase (note that any enzyme ending in -kinase indicates that a phosphate group is going to be donated). Energy is lost because ATP loses a phosphate group and is converted into ADP in the process of creating G6P.

2. G6P is then rearranged, turning from a hexagonal figure to a pentagonal figure, and is configured into Fructose-6-Phosphate (F6P) by an enzyme called phosphoglucose isomerase.

3. F6P is converted into Fructose-1,6-Bisphosphate (F1,6BP) by phosphofructokinase, donating a phosphate group from ATP and turning it into ADP, losing energy.

4. F1,6BP is then broken down into two 3-carbon molecules by the enzyme aldolase. These molecules are DHAP (dihydroxyacetone phosphate) and G3P (glucose-3-phosphate). DHAP acts as a kind of brake on glycolysis, if there is too much energy in the body, this reaction favors DHAP. However, if there is not enough energy in the body, this reaction favors G3P to go on to continue glycolysis.

5. G3P is then converted into 1,3-Bisphosphoglycerate (1,3-BPG) via glyceraldehyde phosphate dehydrogenase. During this conversion, NADP$^+$ is converted into NADPH.

6. 1,3-BPG becomes 3-Phosphoglycerate (3-PG) via the enzyme phosphoglycerate kinase). Energy here is gained in the form of ATP, in which ADP gains a phosphate group through this enzyme.

7. 3-PG is then mutated by phosphoglyceromutase into 2-Phosphoglycerate (2-PG).

8. 2-PG is converted into phosphoenolpyruvate (PEP) via the enzyme enolase. Water is created via this process.

9. Finally, PEP is converted into pyruvate through the enzyme pyruvate kinase, in which energy is again gained by the donation of a phosphate group to ADP to create ATP. Pyruvate is then entered into the Krebs cycle.

The Krebs (Citric Acid) Cycle

If oxygen is present in a eukaryotic organism, the remainder of the process of respiration will occur inside the mitochondria via the Krebs cycle and oxidative phosphorylation. The main goal of the Krebs cycle is to take pyruvate and break it down, producing NADH and $FADH_2$.

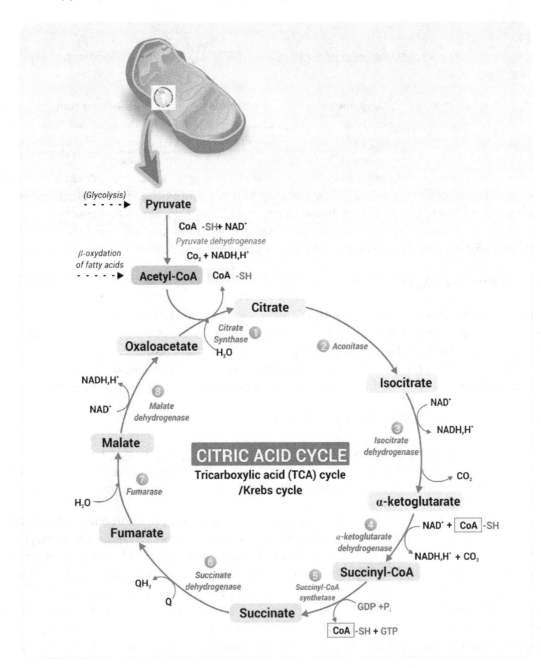

Upon entering the mitochondria, the pyruvate releases a pair of high-energy electrons, and a proton (H^+) to the electron carrier NAD^+, plus a carbon dioxide molecule. This happens while it is being converted into a two-carbon molecule attached to Coenzyme A, formally called Acetyl CoA. This step is called pyruvate oxidation. Acetyl CoA is used as fuel for the following citric acid (Krebs) cycle.

Including the intermediate stage, where pyruvate enters, two spins of the cycle (one for each pyruvate) produces the following:

- 2 CO_2 (from intermediate)
- 2 NADH (from intermediate)
- 4 CO_2
- 6 NADH
- 2 $FADH_2$
- 2 ATP (or GTP) via substrate-level phosphorylation. GTP is guanosine tri-phosphate, which is analogous to ATP.

NADH and $FADH_2$ are electron carriers, meaning they donate electrons to the electron transport chain. NADH donates two high-energy electrons and one proton (H+), while $FADH_2$ donates two electrons and two protons (2 H^+). These high-energy electron and proton carriers then go through the process of oxidative phosphorylation, where many ATP molecules are made. Using the energy supplied by the electron transport chain, protons are transported through the integral protein complexes I, III, and IV along the inner mitochondrial membrane. The pumping of these protons out of the mitochondrial matrix across the cristae to the inner membrane space, establishes a concentration gradient. Just like in photosynthesis, this gradient provides the proton motive force to generate ATP when the hydrogen ion later passes through ATP synthase, causing it to spin and convert ADP to ATP.

Electron Transport Chain

The energy source for oxidative phosphorylation (also known as the electron transport chain) are the electrons traveling through the membrane of the mitochondrial cristae in a redox reaction-driven electron transport chain. Their high energy is coupled with the active transport of the protons, and as they are passed down the chain in a series of redox reactions, oxygen becomes the final electron pair acceptor. The electrons and the hydrogen ions join an electronegative oxygen to form water.

Energy Extraction from Cellular Respiration

Aerobic respiration uses oxygen and produces 30-32 ATP molecules using the mitochondria. Only a few of the ATP are generated via substrate-level phosphorylation in glycolysis and the Krebs cycle; the vast majority of ATP is generated through the electron transport chain and chemiosmosis.

The exact number of ATP molecules made per glucose molecule varies. Glycolysis and the Krebs cycle each produce a net gain of 2 ATP/GTP via substrate-level phosphorylation. Oxidative phosphorylation is more difficult to calculate. Each electron carrier NADH produces around 2.5 ATP, while $FADH_2$ produces around 1.5 ATP. These are not whole numbers because there is not a direct relationship between electron transport and phosphorylation — they are two different processes. One has to do with electrons traveling down the chain to the final electron acceptor: oxygen. The other has to do with the movement of hydrogen ions. Finally, some of the work done by oxidative phosphorylation might be distributed to other cellular processes because respiration does not exist in a vacuum.

Another example of the flexibility of energy production by respiration is seen in thermoregulation. It was discussed earlier that endothermic organisms have ways to regulate body heat, including shivering and sweating. Another is by using an uncoupling protein in the cristae during hibernation. A mitochondrial protein called uncoupling protein 1 (UCP1) in brown fat cells hijacks the proton motive force by preventing them from entering the ATP synthase and creating ATP. Instead, UCP1 moves the protons by

increasing the permeability of the inner mitochondrial membrane, using energy from the proton gradient to be dissipated as heat. This helps keep hibernating animals warm without creating unneeded ATP, which helps animals conserve energy and keep their metabolic rates low.

In addition to two pyruvate molecules produced by glycolysis, six molecules of NADH and two molecules of flavin adenine dinucleotide ($FADH_2$) are produced and used by the ETC. Hydrogen atoms, transported by NADH and $FADH_2$ to the ETC, are used to produce ATP from ADP. The hydrogen atoms form a proton concentration gradient down the ETC that produces energy required to produce ATP. NADH and $FADH_2$ molecules rephosphorylate ADP to ATP via the ETC with each NADH producing three ATP molecules and $FADH_2$ producing two ATP molecules.

Enzymes

Enzymes are a class of catalysts instrumental in biochemical reactions, and in most, if not all, instances are proteins. Like all catalysts, enzymes increase the rate of a chemical reaction by providing an alternate path requiring less activation energy. Enzymes catalyze thousands of chemical reactions in the human body. Enzymes are proteins and possess an active site, which is the part of the molecule that binds the reacting molecule, or substrate. The "lock and key" analogy is used to describe the substrate key fitting precisely into the active site of the enzyme lock to form an enzyme-substrate complex.

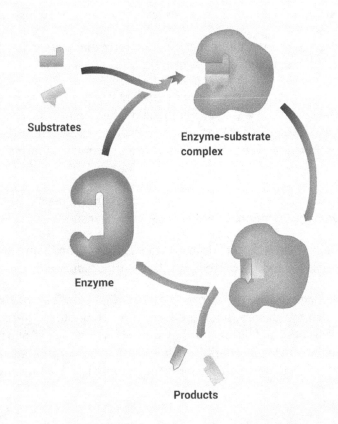

Many enzymes work in tandem with cofactors or coenzymes to catalyze chemical reactions. **Cofactors** can be either inorganic (not containing carbon) or organic (containing carbon). Organic cofactors can be either coenzymes or prosthetic groups tightly bound to an enzyme. **Coenzymes** transport chemical groups from one enzyme to another. Within a cell, coenzymes are continuously regenerating and their concentrations are held at a steady state.

Several factors including temperature, pH, and concentrations of the enzyme and substrate can affect the catalytic activity of an enzyme. For humans, the optimal temperature for peak enzyme activity is approximately body temperature at 98.6 ^{0}F, while the optimal pH for peak enzyme activity is approximately 7-8. Increasing the concentrations of either the enzyme or substrate will also increase the rate of reaction, up to a certain point.

The activity of enzymes can be regulated. One common type of enzyme regulation is termed **feedback inhibition**, which involves the product of the pathway inhibiting the catalytic activity of the enzyme involved in its manufacture.

Biosynthesis

All biological material is made up of atoms bonded together to form molecules. These bonds give the molecules their structure and determine the function of the molecule. Many biologically important molecules are polymers, which are chains of a repeated basic unit, called a monomer.

Carbon is the foundation of organic molecules because it has the ability to form four covalent bonds and long polymers. The four organic compounds are lipids, carbohydrates, proteins, and nucleic acids.

- Lipids are critical for cell membrane structure, long-term energy storage, and to help form some steroid hormones such as testosterone and cholesterol.

- Carbohydrates are important as a medium for energy storage and conversion, but also have structural importance. Cellulose (a polymer made of glucose) provides structure for plant cell walls. Chitin provides structure for fungi and animals with exoskeletons (such as crabs and lobsters), and peptidoglycan is a carbohydrate/protein hybrid that forms the cell walls of some prokaryotes.

- Proteins are important because enzymes regulate all chemical reactions, but there are also many cell membrane proteins important for structure, transport, and communication.

- Nucleic acids include DNA—the genetic instructions of organisms—and RNA, which is the molecule responsible for turning those instructions into products.

All of these organic compounds require the elements carbon (C), hydrogen (H), and oxygen (O), which enter the food chain through the glucose that it produces through photosynthesis. Some compounds contain phosphorus (P), sulfur (S), and nitrogen (N) as well. Elements such as phosphorus and nitrogen diffuse into the roots of plants from the external environment, are incorporated into organic compounds, and are distributed to other organisms through symbiotic relationships or food webs.

Nucleic Acids

Two types of five-carbon sugars can form nucleotides: ribose and deoxyribose. Ribose is the sugar found in ribonucleic acid (RNA). Deoxyribose is the sugar found in deoxyribonucleic acid (DNA), and it has one less oxygen atom than ribose. The important nitrogenous bases are adenine, guanine, cytosine, thymine, and uracil. Thymine is found only in DNA and uracil is found only in RNA. The structure of these molecules determines their function. DNA and RNA have different properties because ribose and

deoxyribose have slightly different structures. DNA is usually found as a double helix because deoxyribose has more flexibility. Nucleic-acid function is also determined by the sequence and properties of the nitrogenous bases. Nitrogenous bases form hydrogen bonds with specific other nitrogenous bases. Adenine interacts with thymine and uracil, and cytosine interacts with guanine; these interactions are called **base pairs**.

This specific pairing allows DNA to serve as the hereditary material for cells because it can be copied accurately and passed down to daughter cells. DNA is also able to serve as hereditary material because the sequence of the nitrogenous bases acts as a code that can be made into all of the proteins needed by the cell. RNA is transcribed from DNA accurately because the nitrogenous bases from RNA interact with those from DNA. Proteins are translated from this RNA using a special type of RNA, called a transfer RNA. The transfer RNA has three nucleotides that can bind to the RNA being used to make the protein.

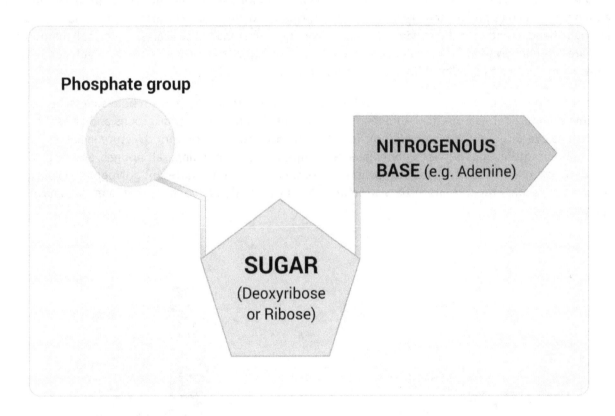

Proteins

Proteins are molecules that consist of carbon, hydrogen, oxygen, nitrogen, and other atoms, and they have a wide array of functions. The monomers that make up proteins are amino acids. All amino acids have the same basic structure. They contain an amino group ($-NH_2$), a carboxylic acid group ($-COOH$), and an R group. The R group, also called the functional group, is different in each amino acid.

The functional groups give the different amino acids their unique chemical properties. There are twenty naturally occurring amino acids that can be divided into groups based on their chemical properties. Glycine, alanine, valine, leucine, isoleucine, methionine, phenylalanine, tryptophan, and proline have nonpolar, hydrophobic functional groups. Serine, threonine, cysteine, tyrosine, asparagine, and glutamine have polar functional groups. Arginine, lysine, and histidine have charged functional groups that are basic, and aspartic acid and glutamic acid have charged functional groups that are acidic.

A **peptide bond** can form between the carboxylic-acid group of one amino acid and the amino group of another amino acid, joining the two amino acids. A long chain of amino acids is called a polypeptide or protein. Because there are so many different amino acids and because they can be arranged in an infinite number of combinations, proteins can have very complex structures. There are four levels of protein structure. Primary structure is the linear sequence of the amino acids; it determines the overall structure of the protein and how the functional groups are positioned in relation to each other, as well as how they interact. Secondary structure is the interaction between different atoms in the backbone chain of the protein. The two main types of secondary structure are the alpha helix and the beta sheet. Alpha helices are formed when the N-H of one amino-acid hydrogen bonds with the C=O of an amino acid four amino acids earlier in the chain.

The functional groups of certain amino acids—including methionine, alanine, uncharged leucine, glutamate, and lysine—make the formation of alpha helices more likely. The functional groups of other amino acids, such as proline and glycine, make the formation of alpha helices less likely. Alpha helices are right-handed and have 3.6 residues per turn. Proteins with alpha helices can span the cell membrane and are often involved in DNA binding. Beta sheets are formed when a protein strand is stretched out, allowing for hydrogen bonding with a neighboring strand.

Similar to alpha helices, certain amino acids have an increased propensity to form beta sheets. Tertiary protein structure forms from the interactions between the different functional groups and gives the protein its overall geometric shape. Interactions that are important for tertiary structure include hydrophobic interactions between nonpolar side groups, hydrogen bonding, salt bridges, and disulfide bonds. Quaternary structure is the interaction that occurs between two different polypeptide chains and involves the same properties as tertiary structure. Only proteins that have more than one chain have quaternary structure.

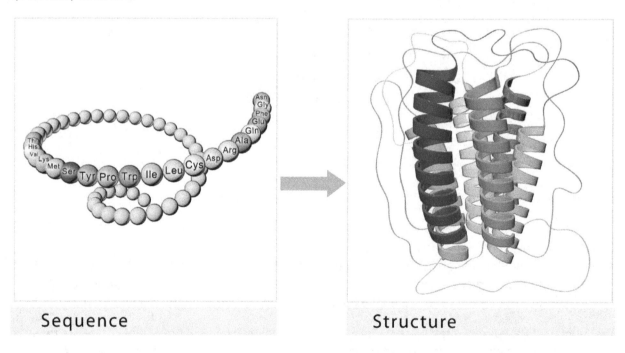

Sequence   Structure

Lipids

Lipids are a very diverse group of molecules that include fats, oils, waxes, and steroids. Since most lipids are primarily nonpolar, they are hydrophobic; however, some lipids have polar regions, making them amphiphilic, which means they are both hydrophobic and hydrophilic. Because the structure of different lipids is so diverse, they have a wide range of functions, which include energy storage, signaling, structure, protection, and making up the cell membrane. Triglycerides are one type of biologically important lipid. They are made up of one molecule of glycerol bonded to three long fatty-acid chains, which are long hydrocarbon chains with a carboxylic acid group.

Many of the properties of triglycerides are determined by the structure of the fatty-acid chains. Saturated triglycerides are made of hydrocarbon chains that only have single bonds between the carbon atoms. Because this structure can pack in more tightly, these triglycerides are often solids at room temperature. Unsaturated triglycerides have double bonds between one or more pairs of carbon atoms. These double bonds create a kink in the fatty-acid chain, which prevents tight packing of the molecules; thus, triglycerides are often liquid at room temperature.

Nucleic acids are polymers whose biological function is to provide hereditary information for all living creatures. The monomer of nucleic acids is a nucleotide. Nucleotides consist of a five-carbon sugar, a nitrogenous base, and a phosphate group.

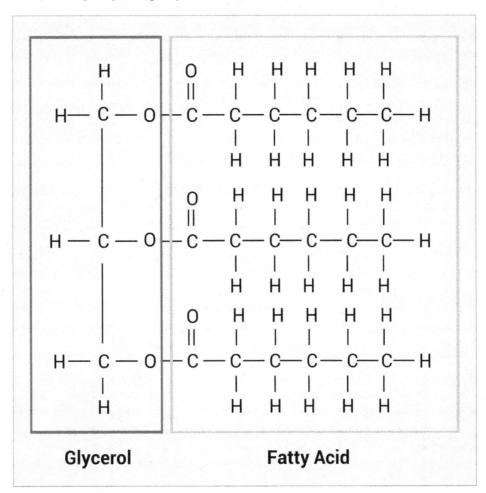

Glycerol   **Fatty Acid**

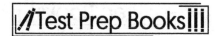

Carbohydrates

Carbohydrates are molecules consisting of carbon, oxygen, and hydrogen atoms and serve many biological functions, such as storing energy and providing structural support. The monomer of a carbohydrate is a sugar monosaccharide. Some biologically important sugars include glucose, galactose, and fructose. These monosaccharides are often found as six-membered rings and can join together by a process known as condensation to form a disaccharide, which is a polymer of two sugar units.

Polysaccharides are long chains of monosaccharides. Their main biological functions are to store energy and provide structure. Starch and cellulose are both polymers of glucose made by plants, but their functions are very different because the structure of the polymers are different. Plants make starch to store the energy that is produced from photosynthesis, while cellulose is an important structural component of the cell wall. These two molecules differ in how the molecules of glucose are bonded together. In starch, the bond between the two glucose molecules is a high-energy alpha bond that is easily hydrolyzed, or broken apart. The bonds between the glucose molecules in cellulose are beta bonds. Being stiff, rigid, and hard to hydrolyze, these bonds give the cell wall structural support. Other important carbohydrates include glycogen, which provides energy storage in animal cells, and chitin, which provides structure to arthropod exoskeletons.

Directionality Affects the Structure and Function of a Polymer

The structure of these biological molecules often results in directionality, where one end is different from the other.

In nucleic acids, the five carbons that make up the sugar molecule are numbered. The 5' carbon is bound to the phosphate group and the 3' carbon is bound to a hydroxyl group (-OH). A nucleotide strand can only grow by bonding the hydroxyl group to the phosphate group, so one end of nucleic acids is always a 5' end and the other is always a 3' end. New nucleotides can only be added to the 3' end of the nucleic-acid strand. Therefore, replication and transcription only occur in the 5' to 3' direction.

Proteins also have directionality. One end of an amino acid contains the NH group and the other end contains the COOH group. This structure gives the amino acids and peptides directionality. A peptide bond can only form between the NH group of one amino acid and the peptide group of another.

Carbohydrates consist of sugars and polymers of sugars. The simplest sugars are monosaccharides, which have the empirical formula of CH_2O. The formula for the monosaccharide glucose, for example, is $C_6H_{12}O_6$. Glucose is an important molecule for cellular respiration, the process of cells extracting energy by breaking bonds through a series of reactions. The individual atoms are then used to rebuild new small molecules. It is important to note that monosaccharides are the smallest functional carbohydrates, so they cannot be hydrolyzed into simpler molecules and still remain carbohydrates.

Polysaccharides are made up of a few hundred to a few thousand monosaccharides linked together. These larger molecules have two major functions. The first is that they can function as storage molecules, such as starch or glycogen, and then broken down later for energy. Secondly, they may be used to form strong materials, such as cellulose, which is the firm wall that encloses plant cells, and chitin, the carbohydrate insects use to build exoskeletons.

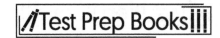

Biological Chemistry

Biochemical Processes

Life processes require energy. Energy is defined as the ability to do work or move matter against external forces. Free energy is the energy within a physical system that is used to do work. There are two basic forms of energy: potential and kinetic. **Potential energy** is the energy stored within an object that has the capacity to do work, but hasn't yet. For example, consider two equally strong people pulling a rope on opposite ends as hard as they can. While they are in this position, nothing is happening. However, energy stored as tension in the rope has the *potential* to do work. If one person suddenly let go of the rope, the person on the other end would fall backward. Once this person falls, the potential energy is now converted into **kinetic energy**—the energy of movement. Potential energy is the most important type of energy in biochemical reactions, since high amounts of potential energy stored in the atomic bonds of molecules can be metabolized to release kinetic energy, which can be harnessed to do many different kinds of work.

Entropy and Maintaining Homeostasis

In order for organisms to survive against this universal tendency for chaos, they must use energy to work against entropy. They do this via biochemical processes that maintain an internal order, called homeostasis. Homeostasis is the physiological processes within a system that regulate a stable internal equilibrium such as body temperature, blood pH, and fluid balance.

Organisms maintain homeostasis by using free energy and matter, usually in the form of atoms and molecules. Atoms are the smallest whole units of matter, and molecules are atoms glued to each other via chemical bonds. Everything is made up of atoms, even the most complicated structures; their complexity is only due to the variety and quantity of different molecular arrangements. A famous experiment by Stanley Miller and Harold Urey simulated early Earth's atmospheric conditions and witnessed the synthesis of a string of molecules known as amino acids. Amino acids are the blueprints of proteins — the building blocks of life. This famous experiment provided a glimpse of the original atomic arrangements that facilitated the origin of life. Before proteins, random particles would collide, due to chance, only if given enough time. After proteins arrived, life arose, likely because proteins enable organisms to use available energy and matter to make complicated structures, from cells to whole organisms. Enzymes (proteins that act as catalysts) help reactants find each other by providing a docking station, increasing probabilities of atomic collision and bonding, and therefore, facilitate the specific biochemical reactions that make life possible.

Here's an illustration of that:

The Lock and Key Mechanism

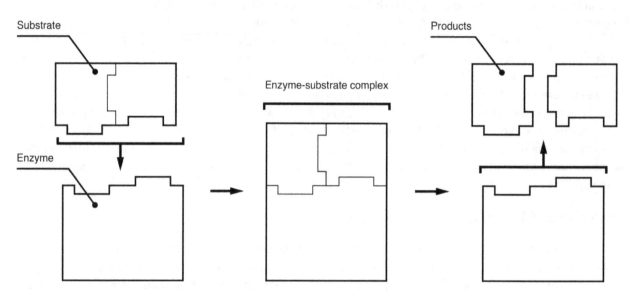

Proteins combine to make a myriad of different structures. Think of proteins like people. A pile of bricks can't do anything on its own. Buildings can only form when there are workers to move the bricks into place. Similarly, proteins turn matter into the complicated structures that form organisms.

Proteins in the form of enzymes are especially important for biochemical reactions because they lower the activation energy—the minimum energy required for a chemical reaction to happen—as illustrated in the below graph:

Using available free energy, reactions occur within an organism that create organization and decrease entropy. The below graphs illustrate how free energy is used in two basic biochemical reactions:

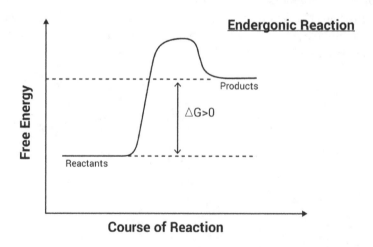

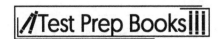

The image above illustrates a reaction that decreases entropy (or increases order), called an **endergonic reaction**. Endergonic, or **anabolic**, reactions occur when reactants *absorb* energy from the surroundings so that the products hold more energy than the reactants. Anabolic reactions enable the organism to make bigger things (polymers) from smaller things (monomers), such as forming new cells, or making proteins from amino acids.

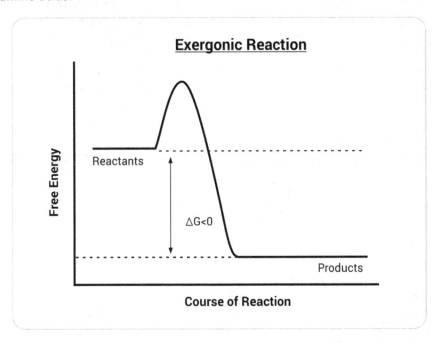

Conversely, the image above illustrates an **exergonic**, or **catabolic**, reaction that *releases* energy. In this reaction, the products hold less energy than the reactants, and entropy (or disorder) is increased. Catabolic reactions enable organisms to break down bigger things (polymers) into smaller things (monomers); for example, breaking down proteins into their respective amino acids.

As long as the energy released (exergonic) exceeds the energy absorbed (endergonic) in an organism, it will continue to function and sustain life. If there is not energy available from exergonic reactions to drive endergonic reactions, the organism will die, since energy is required to perform all metabolic functions. Metabolism is the set of processes carried out by an organism that permits the exchange of energy between itself and the environment, enabling it to change and grow internally.

In other words: metabolism = catabolism + anabolism

- If catabolism is greater than anabolism, i.e., energy released is greater than energy consumed, then an organism can live.

- If catabolism is less than anabolism, i.e., energy released is less than energy consumed, then an organism will die.

Organisms must consume energy, in the form of food or light, to perform anabolic reactions. If they are autotrophs, such as plants, they produce their own food. If they are heterotrophs, such as animals, they absorb the energy provided by food, most commonly sugars. The most common form of sugar used for energy is the simple sugar glucose, $C_6H_{12}O_6$, which contains extremely high amounts of potential energy stored within its atomic bonds.

ATP

As mentioned, cellular respiration is the catabolic process of breaking down the bonds in glucose and releasing its potential energy in the form of ATP, or adenosine triphosphate. ATP harnesses small amounts of energy and uses it for processes in cellular metabolism. Each glucose molecule can produce about 32 ATP molecules. Breaking glucose and storing its energy in smaller molecules enables the cells to distribute energy across many metabolic reactions instead of just one.

ATP holds energy in the bonds between its phosphates. It also cycles back and forth between harnessing and distributing energy by forming and breaking a phosphate bond, as shown in the figure below.

The ATP - ADP Cycle

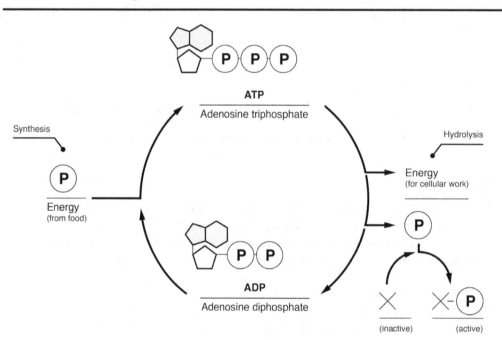

Exergonic Reactions Can Drive Endergonic Reactions

Cells balance their energy resources by using the energy from exergonic reactions to drive endergonic reactions forward, a process called energy coupling. Adenosine triphosphate, or ATP, is a molecule that is an immediate source of energy for cellular work. When it is broken down, it releases energy used in endergonic reactions and anabolic pathways. ATP breaks down into adenosine diphosphate, or ADP, and a separate phosphate group, releasing energy in an exergonic reaction. As ATP is used up by reactions, it is also regenerated by having a new phosphate group added onto the ADP products within the cell in an endergonic reaction.

Even within a single process, there are energy investments and divestments in many necessary reactions along the way. ATP is necessary for all of these reactions. It can be made in two different ways: substrate-level phosphorylation and oxidative phosphorylation.

Substrate-level phosphorylation occurs when a phosphoryl group (PO_3^{2-}) is donated to an ADP (adenosine diphosphate) molecule from a substrate by using enzymes, thereby creating ATP, as shown below in the image on the left. In oxidative phosphorylation, as illustrated on the right, a free phosphate joins ADP via the energy provided by chemiosmosis and the mechanical movement of the ATP synthase (the enzyme that creates ATP).

Obtaining energy at the cellular level and understanding how energy investment is used to create organization in living things, is important in order to understand energy exchange between all of the cells of an organism and even between an organism and its environment.

Practice Questions

1. How do cellulose and starch differ?
   a. Cellulose and starch are proteins with different R groups.
   b. Cellulose is a polysaccharide made up of glucose molecules and starch is a polysaccharide made up of galactose molecules.
   c. Cellulose and starch are both polysaccharides made up of glucose molecules, but they are connected with different types of bonds.
   d. Cellulose and starch are the same molecule, but cellulose is made by plants and starch is made by animals.
   e. Cellulose and starch are both polysaccharides made up of glucose molecules, but starch is insoluble and cellulose is soluble.

2. A mutation in the sequence of a protein causes the secondary structure to change. How did the mutation cause the change?
   I. The R group from the mutated amino acid interacts differently with other R groups.
   II. The R group from the mutated amino acid prevents the formation of hydrogen bonds between the atoms of the backbone of the protein.
   III. The mutation causes peptide bonds to change.

   a. Choice I only
   b. Choice II only
   c. Choice III only
   d. Choices I and II
   e. Choices I and III

3. Palmitoleic acid is a fatty acid with one double bond in the hydrocarbon chain. What property would you expect from palmitoleic acid?
   a. Solid at room temperature
   b. Gas at room temperature
   c. Liquid at room temperature
   d. Hydrophilic
   e. Miscible in water

4. What molecule serves as the hereditary material for prokaryotic and eukaryotic cells?
   a. Proteins
   b. Carbohydrates
   c. Lipids
   d. DNA
   e. RNA

5. What organelles have two layers of membranes?
   a. Nucleus, ER, mitochondria
   b. Nucleus, Golgi apparatus, mitochondria
   c. ER, chloroplast, lysosome
   d. Chloroplast, lysosome, ER
   e. Nucleus, chloroplast, mitochondria

6. What organelle is the site of protein synthesis?
   a. Nucleus
   b. Smooth ER
   c. Ribosome
   d. Lysosome
   e. Golgi apparatus

7. Certain bacteria that can break down the bonds in cellulose live in the gut of ruminants, which are mammals that feed primarily on grasses. Animals cannot break down cellulose. How does this affect the energy efficiency of both the bacteria and the ruminants?
   a. Energy efficiency of the bacteria increases. Energy efficiency of the ruminants decreases.
   b. Energy efficiency of the bacteria decreases. Energy efficiency of the ruminants decreases.
   c. Energy efficiency of the bacteria decreases. Energy efficiency of the ruminants increases.
   d. Energy efficiency of the bacteria increases. Energy efficiency of the ruminants increases.
   e. Energy efficiency does not change for either the bacteria nor the ruminants.

8. Two different bacterial cultures are grown from bacteria with the same genome sequence. Transcriptional analysis shows that Culture B is expressing genes that can metabolize lactose, but Culture A is not. How can this happen if they have the same genetic sequence?
   I. Someone mislabeled the tubes and the bacteria must have different genome sequences.
   II. Culture A is grown in the presence of lactose, which turns on a different set of genes.
   III. Culture B is grown in the presence of lactose, which turns on a different set of genes.

   a. I only
   b. II only
   c. III only
   d. II and III
   e. I and II

9. Which of the following structures is unique to eukaryotic cells?
   a. Cell walls
   b. Nuclei
   c. Cell membranes
   d. Organelles
   e. Ribosomes

10. Which is the cellular organelle used for digestion to recycle materials?
    a. The Golgi apparatus
    b. The lysosome
    c. The centrioles
    d. The mitochondria
    e. The ribosome

11. Which base pairs with adenine in RNA?
    a. Thymine
    b. Guanine
    c. Cytosine
    d. Uracil
    e. It depends on whether it is tRNA or mRNA

12. Both NADP⁺ and NAD⁺ are important for cellular energy conversion. They distribute high-energy electrons to electron transport chains that facilitate the pumping of protons across membranes and couple the action with redox reactions. Which of the following is another similarity between the two molecules?

    a. They both generate ATP by traveling through ATP synthase.
    b. They both carry one proton and a pair of electrons when they are reduced.
    c. They both deliver electrons from either glycolysis or the Krebs cycle to the cristae.
    d. They both are oxidized by electrons energized by the photosystems.
    e. The ratio of [NADP⁺]/[NADPH] and [NAD⁺]/[NADH] in the cell are the same.

13. Which of the following is true about an endergonic reaction?

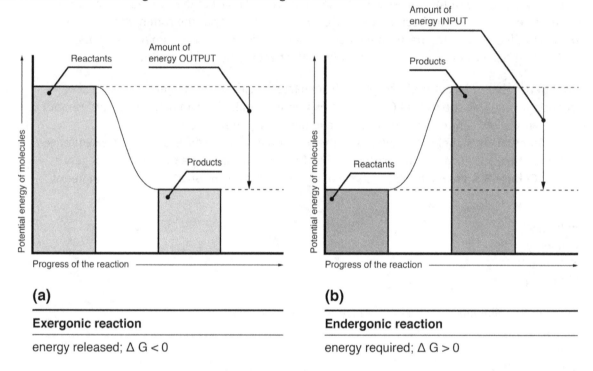

(a)

Exergonic reaction

energy released; Δ G < 0

(b)

Endergonic reaction

energy required; Δ G > 0

    a. The reaction releases energy and decreases entropy.
    b. The reaction absorbs energy and increases entropy.
    c. The reaction releases energy and increases entropy.
    d. The reaction absorbs energy and decreases entropy.
    e. The reaction releases energy but does not change entropy.

14. What is the term used for the set of metabolic reactions that convert chemical bonds to energy in the form of ATP?

    a. Photosynthesis
    b. Reproduction
    c. Active transport
    d. Energy expenditure
    e. Cellular respiration

15. Which of the following is NOT true regarding cell cycle checkpoints in mitosis?
    a. A cyclin/CDK pair is responsible for assembling mitotic machinery.
    b. Cyclin protein increases in interphase and is broken down during mitosis.
    c. CDK protein is equally expressed throughout the cell cycle.
    d. G_0 is a state that stimulates progression into S phase.
    e. The G_1 checkpoint prepares and commits cells to entering the cell cycle.

16. What types of molecules can move through a cell membrane by passive transport?
    a. Complex sugars
    b. Non-lipid soluble molecules
    c. Oxygen
    d. Molecules moving from areas of low concentration to areas of high concentration
    e. Amino acids

17. What is ONE feature that both prokaryotes and eukaryotes have in common?
    a. A plasma membrane
    b. A nucleus enclosed by a membrane
    c. Organelles
    d. A nucleoid
    e. Linear DNA

18. What is the LAST phase of mitosis?
    a. Prophase
    b. Telophase
    c. Anaphase
    d. Metaphase
    e. Prometaphase

19. In which organelle do eukaryotes carry out aerobic respiration?
    a. Golgi apparatus
    b. Nucleus
    c. Mitochondrion
    d. Cytosol
    e. Chloroplast

20. What kind of energy do plants use in photosynthesis to create chemical energy?
    a. Chemical
    b. Electric
    c. Nuclear
    d. Cellular
    e. Light

21. What type of biological molecule includes monosaccharides?
    a. Protein
    b. Carbohydrate
    c. Nucleic acid
    d. Lipid
    e. Enzyme

22. Which level of protein structure is defined by the folds and coils of the protein's polypeptide backbone?
    a. The amino acid sequence
    b. Primary
    c. Secondary
    d. Tertiary
    e. Quaternary

Answer Explanations

1. C: Cellulose and starch are both polysaccharides that are long chains of glucose molecules, but they are connected by different types of bonds, which gives them different structures and different functions.

2. B: Secondary structure is formed from hydrogen bonds between the backbone atoms in the protein chain. Some R groups allow these interactions to form, and others prevent them. A mutation can cause the secondary structure to change when it changes an R group that allows these interactions to one that prevents these interactions.

3. C: Palmitoleic acid is an unsaturated fatty acid. They are typically liquids at room temperature. They are also hydrophobic. Most oils are immiscible in water.

4. D: DNA serves as the hereditary material for prokaryotic and eukaryotic cells.

5. E: The nucleus, chloroplast, and mitochondria are all bound by two layers of membrane. The Golgi apparatus, lysosome, and ER only have one membrane layer.

6. C: Proteins are synthesized on ribosomes. The ribosome uses messenger RNA as a template and transfer RNA brings amino acids to the ribosome where they are synthesized into peptide strands using the genetic code provided by the messenger RNA.

7. D: The ruminants provide a food sources for the bacteria, and the bacteria help the ruminants utilize their main food source. Therefore, the energy efficiency of both organisms increases.

8. C: Gene expression can be influenced by the environment. Lactose metabolism is regulated by the presence of lactose. Bacteria that have the genes to metabolize lactose will turn them off if lactose is not present but will turn them on if lactose is present.

9. B: The structure exclusively found in eukaryotic cells is the nucleus. Animal, plant, fungi, and protist cells are all eukaryotic. DNA is contained within the nucleus of eukaryotic cells, and they also have membrane-bound organelles that perform complex intracellular metabolic activities. Prokaryotic cells (archaea and bacteria) do not have a nucleus or other membrane-bound organelles and are less complex than eukaryotic cells.

10. B: The cell structure responsible for cellular storage, digestion and waste removal is the lysosome. Lysosomes are like recycle bins. They are filled with digestive enzymes that facilitate catabolic reactions to regenerate monomers. The Golgi apparatus is designed to tag, package, and ship out proteins destined for other cells or locations. The centrioles typically play a large role only in cell division when they ratchet the chromosomes from the mitotic plate to the poles of the cell. The mitochondria are involved in energy production and are the powerhouses of the cell. Ribosomes are the sites of protein synthesis.

11. D: DNA and RNA each contain four nitrogenous bases, three of which they have in common: adenine, guanine, and cytosine. Thymine is only found in DNA, and uracil is only found in RNA. Adenine interacts with uracil in RNA, and with thymine in DNA. Guanine always pairs with cytosine in both DNA and RNA. Base pairing is the same in the synthesis of tRNA and mRNA.

12. B: $NADP^+$ and NAD^+ are very similar molecules in that they carry a proton and steal a pair of high-energy electrons when they are reduced to NADPH and NADH. The major difference is that $NADP^+$ is the electron carrier in photosynthesis and NAD^+ is the electron carrier in respiration.

$NADP^+$, not NAD^+, takes electrons from the electrons excited by Photosystem I, making Choice D incorrect, not to mention that they are not oxidized. NAD^+, not $NADP^+$, is responsible for reducing high-energy glucose derivatives in glycolysis and in the Krebs cycle, making Choice *C* incorrect.

The electrons of reduced $NADP^+$ are invested in sugars in the Calvin cycle, and the electrons from reduced NAD^+ are delivered to an electron transport chain in the cristae. Neither molecule travels through ATP synthase, making Choice *A* incorrect.

Choice *E* is incorrect because the ratio of $[NADP^+]/[NADPH]$ and $[NAD^+]/[NADH]$ in the cell are not the same. The ratio of $[NADP^+]/[NADPH]$ is around 1,000 and the ratio of $[NAD^+]/[NADH]$ is about 0.01.

13. D: An endergonic reaction absorbs energy to make bigger things from smaller things, resulting in an increase in order, or a decrease in entropy, because the molecules become condensed into a more rigid form. Choices *A* and *E* are incorrect because it does not release energy, it absorbs it. Choice *B* is incorrect because it does not increase entropy. Choice *C* is incorrect because it does not release energy and increase entropy—these are the requirements of an exergonic reaction.

14. E: Cellular respiration is the term used for the set of metabolic reactions that convert chemical bonds to energy in the form of ATP. All respiration starts with glycolysis in the cytoplasm, and in the presence of oxygen, the process will continue to the mitochondria. In a series of oxidation/reduction reactions, primarily glucose will be broken down so that the energy contained within its bonds can be transferred to the smaller ATP molecules. It's like having a $100 bill (glucose) as opposed to having one hundred $1 bills. This is beneficial to the organism because it allows energy to be distributed throughout the cell very easily in smaller packets of energy.

15. D: G_0 is a state that stimulates progression into S phase. G_0 is actually a checkpoint that involves cells exiting the cell cycle, such as mature neurons and damaged cells that may undergo apoptosis. Therefore, Choice *D* is the untrue statement. Choices *A*, *B*, and *C* refer to the Maturation Promoting Factor (MPF), which is a cyclin/CDK pair that forms in G_2 when rising levels of cyclin bind to and activate an ever-present cyclin dependent kinase. Choice *E* is also true as the G_1 checkpoint serves to prepare and commit cells to entering the cell cycle.

16. C: Molecules that are soluble in lipids, like fats, sterols, and vitamins (A, D, E and K), for example, are able to move in and out of a cell using passive transport. Water and oxygen are also able to move in and out of the cell without the use of cellular energy. Complex sugars, amino acids, and non-lipid soluble molecules are too large to move through the cell membrane without relying on active transport mechanisms. Molecules naturally move from areas of high concentration to those of lower concentration. It requires active transport to move molecules in the opposite direction, as suggested by Choice *D*.

17. A: Both types of cells are enclosed by a cell membrane, which is selectively permeable. Selective permeability means essentially that it is a gatekeeper, allowing certain molecules and ions in and out, and keeping unwanted ones at bay, at least until they are ready for use. Prokaryotes contain a nucleoid and do not have organelles; eukaryotes contain a nucleus enclosed by a membrane, as well as organelles. DNA in prokaryotes is circular, while it is linear in eukaryotes.

18. B: During telophase, two nuclei form at each end of the cell and nuclear envelopes begin to form around each nucleus. The nucleoli reappear, and the chromosomes become less compact. The microtubules are broken down by the cell, and mitosis is complete. The process begins with prophase as the mitotic spindles begin to form from centrosomes. Prometaphase follows, with the breakdown of the nuclear envelope and the further condensing of the chromosomes. Next, metaphase occurs when the microtubules are stretched across the cell and the chromosomes align at the metaphase plate. Finally, in the last step before telophase, anaphase occurs as the sister chromatids break apart and form chromosomes.

19. C: The mitochondrion is often called the powerhouse of the cell and is one of the most important structures for maintaining regular cell function. It is where aerobic cellular respiration occurs and where most of the cell's ATP is generated. The number of mitochondria in a cell varies greatly from organism to organism and from cell to cell. Cells that require more energy, like muscle cells, have more mitochondria. The chloroplast is the site of photosynthesis in plants.

20. E: Photosynthesis is the process of converting light energy into chemical energy, which is then stored in sugar and other organic molecules. The photosynthetic process takes place in the thylakoids inside chloroplast in plants. Chlorophyll is a green pigment that lives in the thylakoid membranes and absorbs photons from light.

21. B: Sugars are carbohydrates. The simplest sugar molecule is called a monosaccharide and has the molecular formula of CH_2O, or a multiple of that formula. Monosaccharides are important molecules for cellular respiration. Their carbon skeleton can also be used to rebuild new small molecules. Lipids include fats, proteins are formed via amino acids, and DNA and RNA are nucleic acids. Enzymes are a type of protein.

22. C: The secondary structure of a protein refers to the folds and coils that are formed by hydrogen bonding between the slightly charged atoms of the polypeptide backbone. The primary structure is the sequence of amino acids (so Choices *A* and *B* are technically the same and both incorrect), similar to the letters in a long word. The tertiary structure is the overall shape of the molecule that results from the interactions between the side chains that are linked to the polypeptide backbone. The quaternary structure is the complete protein structure that occurs when a protein is made up of two or more polypeptide chains.

Ecology

Energy Flow

Ecosystems are maintained by cycling the energy and nutrients that they obtain from external sources. The process can be diagramed in a **food web**, which represents the feeding relationship between species in a community. The different levels of the food web are called **trophic levels**. The first trophic level generally consists of plants, algae, and bacteria. The second trophic level consists of herbivores. The third trophic level consists of predators that eat herbivores. The trophic levels continue on to larger and larger predators. **Decomposers** are an important part of the food chain that are not at a specific trophic level. They eat decomposing things on the ground that other animals do not want to eat. This allows them to provide nutrients to their own predators.

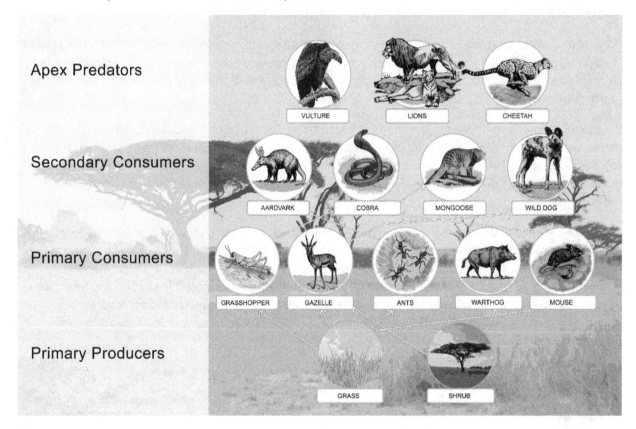

Energy flows from the primary producers to consumers, as consumers eat producers and other consumers. From one step to the next, 90% of the energy is lost. Therefore, in order for a tertiary consumer to receive just 1 kcal of energy, a producer needs to produce 1000 kcal of energy. The energy flow in a community can be shown as a pyramid that illustrates the loss of energy or as a web that illustrates the complex relationships between each organism.

It is important to note that the start of every single food web has a producer. Without a producer to harness the energy from the sun, the entire web would collapse. There would not be any energy to sustain any populations in the community.

Nutrient Cycles

Organisms depend on energy and material for nutrients and survival. Material resources are finite and must circulate within the ecosystem, creating complex recycling systems. Four important resources have well-established cycles:

Water: The sun heats bodies of water and causes evaporation, the changing of water from a liquid to a gas. Plants release water through transpiration, which is when water moves upward through the plant and is then released as a vapor from the leaves into the atmosphere. The cooler temperatures of the atmosphere cause the water to condense back into a liquid, forming a cloud. Eventually, the water in the clouds falls from the sky as precipitation, such as rain, snow, or hail. It then falls back into bodies of water or onto land. On land, precipitation is absorbed into the soil, where it finds its way into bodies of water as runoff or is absorbed into plants.

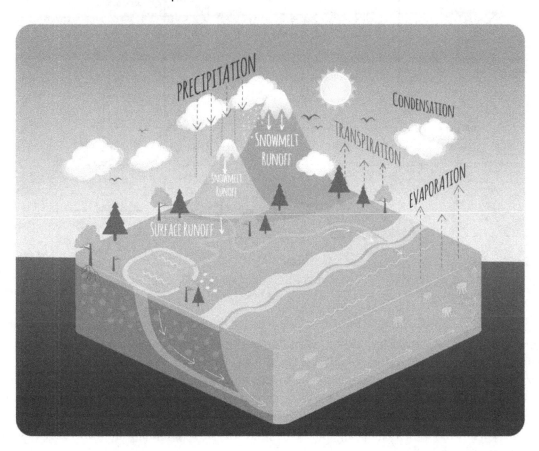

Carbon cycle: Carbon forms the backbone of all biologically important molecules. It is found in the atmosphere as CO_2. Plants, algae, and cyanobacteria take CO_2 and make carbohydrates during photosynthesis using energy from the sun. The carbon then moves through animals in the food chain and is returned to the atmosphere as CO_s during respiration. Decaying biological material, called *detritus,* also provides carbon to the soil. A final source of carbon is the burning of wood and fossil fuels, which releases CO_2 into the atmosphere.

The Carbon Cycle

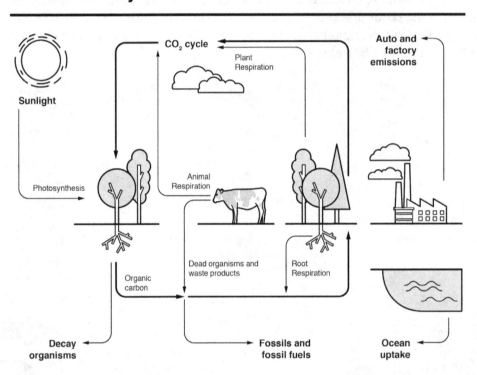

Nitrogen: Nitrogen is important for protein synthesis. Most nitrogen is found in the atmosphere as N_2, which is unusable to the majority of organisms. Lightning can convert N_2 to nitrates, which can then be used by plants. Nitrogen-fixing bacteria can turn N_2 into nitrates or ammonium. These bacteria can live in the roots of legumes, allowing the nitrates to be consumed as food. Ammonium can also be produced by decomposing material. Ammonium is converted into nitrites and then nitrites are converted to nitrates by nitrifying bacteria. Plants can then take up nitrates by assimilation. Denitrifying bacteria convert nitrates back into N_2.

The nitrogen cycle:

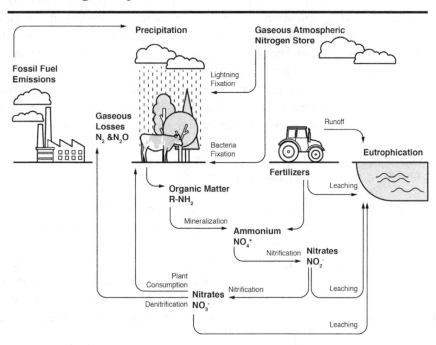

Phosphorous cycle: Phosphorous is important for plant growth and development. It is a component of RNA, DNA, ATP, cell membranes, and bones; and is an important modifier of proteins. Phosphorous is found in rocks and soil as phosphate (PO_4^{3-}), which enters the soil from rock erosion or enters lakes from runoff. PO_4^{3-} is then taken up by plants and algae, which can convert the PO_4^{3-} into useable material. Animals receive phosphorous by eating plants. Phosphorous is returned to the soil after the death of animals and plants. Phosphorous can also enter the soil in the form of man-made fertilizer.

Unlike these recyclable resources, energy flows from sunlight to primary producers to consumers and is not recycled. Between each step, there is actually a loss of energy to the environment, biological processes, and heat generation. Primary producers are the first step in harnessing the energy from sunlight by creating glucose from CO_2 and sunlight. This reaction is greatly affected by changes to the environment. Since light is a necessary factor for photosynthesis, the seasonal variation in daylight affects the rate of photosynthesis. Temperature is also a factor. Photosynthesis slows at low temperatures and transpiration increases at high temperatures, creating water loss. Under drought conditions or high temperatures, photosynthesis occurs less. Photosynthesis increases as the levels of CO_2 increase. Since rates of photosynthesis are greatly affected by these resources, climate change has an impact on photosynthesis. The increasing CO_2 levels impact photosynthesis, as do drastic changes in weather patterns, such as temperature shifts and droughts.

Populations

A population consists of all of the members of the same species that live in a geographical area and are able to reproduce and create fertile offspring. The size of any given population is determined by many factors, which include birth rate, death rate, and migration.

Population Size

The size of a population is affected by many different factors in a community. Populations grow when new members are born, and they shrink when members die or migrate away. Birth rates, death rates, and migratory rates are affected by factors such as the presence of predators, the availability of food, and the availability of shelter. The population size of predators and prey are linked. The population of a predator grows when the number of prey is great. Eventually, the population of the predator gets so high that competition exists between individuals of the predator species. The population of the prey starts to fall and the competition between the predators gets worse. Eventually, the population of the predator cannot be sustained and the population starts to fall. When the predator population decreases enough, the population of the prey starts to rebound. This relationship can also be affected by external factors. Mathematical modeling can predict this relationship and show the impact of the external factors.

Mathematical Models of Population Growth

Mathematical population models can predict the impact of the community on population growth. In a community where a population has unlimited access to all resources, population growth will be exponential. This is the maximum growth rate that a population can have and is not frequently seen in nature. When the population size exceeds the resources, the growth rate will change because individuals will be competing for resources. Birth rates will slow, while death and migratory rates will rise. The population will enter the logistic growth phase and reach the maximum that the community can carry. Assuming there are no changes to the community, the population will reach a steady state.

The human population growth rate can be predicted by fecundity, or reproductive rate, and the age of the population. For the last 100 years, the global human population growth rate has been very high, but, over the past few years, it has started to slow. In certain areas of the world, the growth rate remains high. These places have a young population and a high birth rate. In a young population, there are many people who will still have children and a high birth rate means that a lot of children are born. Other places have a lower birth rate and an older population. In an older population, there are fewer people with reproductive potential and, generally, there is a higher death rate. With a lower birth rate, there are fewer children born to replace the people that die. Almost every part of the world still has an increasing human population growth rate. An exception to this is Japan, where the birth rate is very low and the population is very old and possibly shrinking. Models predict that is may soon happen in Western Europe as well.

Growth Curves and Carrying Capacity

Population dynamics can be characterized by **growth curves**. Growth can either be **unrestricted**, which is modeled by an exponential curve, or **restricted**, which is modeled by a logistic curve. Population growth can be restricted by environmental factors such as the availability of food and water sources,

habitat, and other necessities. The **carrying capacity** of a population is the maximum population size that an environment can sustain indefinitely, given all of the above factors.

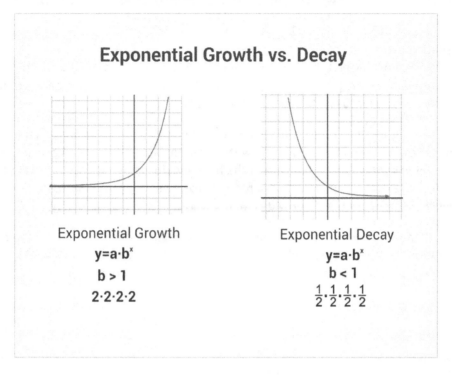

Communities

A community is made up of all of the different populations that live in a given geographic area. There are complex interactions both within a species and between different species. Communities that are more diverse and more complex are more stable than simple communities.

An **ecological community** is a group of species that interact and live in the same location. Because of their shared environment, they tend to have a large influence on each other.

Niche

An **ecological niche** is the role that a species plays in its environment, including how it finds its food and shelter. It could be a predator of a different species, or prey for a larger species.

Species Diversity

Species diversity is the number of different species that cohabitate in an ecological community. It has two different facets: **species richness**, which is the general number of species, and **species evenness**, which accounts for the population size of each species.

Interspecific Relationships

Different populations interact to create the complex functions of a community. These interactions can have both positive and negative impacts on the individuals involved from the different populations and can be modeled mathematically. The models can demonstrate how the negative or positive relationship will impact the population of each species in the relationship. There are five types of interactions:

Competition: Competition is when two individuals vie for a finite amount of resources, such as food, water, and mates. This can occur within or between species. Both groups are negatively affected by competition.

Predation: Predation occurs when one species, the predator, feeds on another species, the prey. Predation usually, but not always, ends in the death of the prey. This relationship is beneficial to the predator and harmful to the prey. Ultimately though, as the number of prey decrease (from predation), the number of predators eventually decrease, as food supplies dwindle.

Parasitism: Parasitism is another relationship where one species, the parasite, gains a benefit from the relationship, but the other, the host, is harmed by the relationship. Unlike predation, however, parasitism does not always result in the death of the host and does not always involve a way to ascertain food. Parasites can use their host as a food source, but they can also use their hosts as a place to lay their eggs for reproduction or to provide a habitat.

Commensalism: Commensalism is when one member of the relationship benefits and the other is not affected at all. Examples of this include when one organism uses another for transport or housing without harming the other organism.

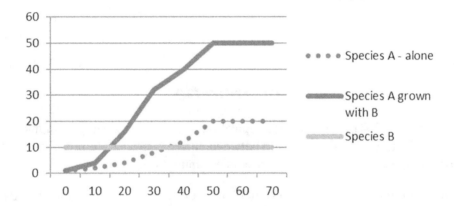

Mutualism: Mutualism is a relationship where both members benefit. There are many examples of this, including: the nectar-drinking/pollination relationship between insects or birds and plants; animals eating fruit and dispersing seeds; and animals that feed on parasites on other animals.

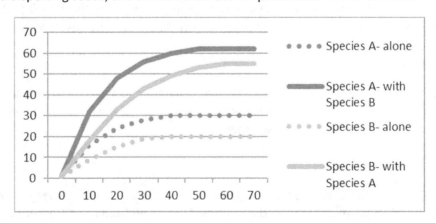

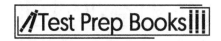

All of these relationships occur in the context of an entire ecosystem. They are not solitary relationships and are affected greatly by other forces outside of the relationship. This allows for feedback mechanisms to control the symbiotic relationships.

Population Dynamics

Population dynamics can be influenced by many factors, both biotic and abiotic. Natural and man-made disasters can greatly affect population size and species distribution. One example of this is the destruction of elm trees by Dutch elm disease. This fungus is spread by beetles and was accidentally introduced to Europe, North America, and New Zealand. The disease is native to Asia and many trees there are resistant to the disease. Dutch elm disease wiped out approximately 75% of the elm trees through much of Europe.

Ecosystems

An **ecosystem** is the basic unit of ecology and combines all of the living, or biotic, organisms and nonliving, or abiotic, components. The abiotic components include light, water, air, minerals, and nitrogen. They provide necessary nutrients and energy for the biotic components. Energy in an ecosystem usually flows from sunlight to primary producers, which are organisms that can undergo photosynthesis, to secondary consumers, and finally to decomposers. Abiotic nutrients often have complex cycling systems between different members of an ecosystem.

Distribution of Ecosystems

While ecosystems are often discussed in their current state, it is important to note that ecosystems are not static; they change over time, which can be a natural process. Weather patterns, geological events, and fires are among the things that can affect an ecosystem. These items can reduce or increase resources, destroy habitats, or even kill individuals. One example of a natural change to an ecosystem is the weather pattern known as El Niño, which is caused by increased water temperatures in the Pacific Ocean and results in weather changes throughout the world. El Niño can have an effect on many ecosystems. In the ocean off the coast of Peru, there are fewer predatory fish because the ocean conditions provide fewer nutrients for the plankton that the fish eat. In areas that experience flooding from El Niño, there can be overflow of salt water into fresh-water systems, which is destructive to those ecosystems. Areas with droughts can see a loss of producers, which is also destructive to ecosystems.

Biomes

A **biome** is a group of plants and animals that are found in many different continents and have the same characteristics because of the similar climates in which they live. Each biome is composed of all of the ecosystems in that area. Five primary types of biomes are aquatic, deserts, forests, grasslands, and tundra. The sum total of all biomes comprises the Earth's biosphere.

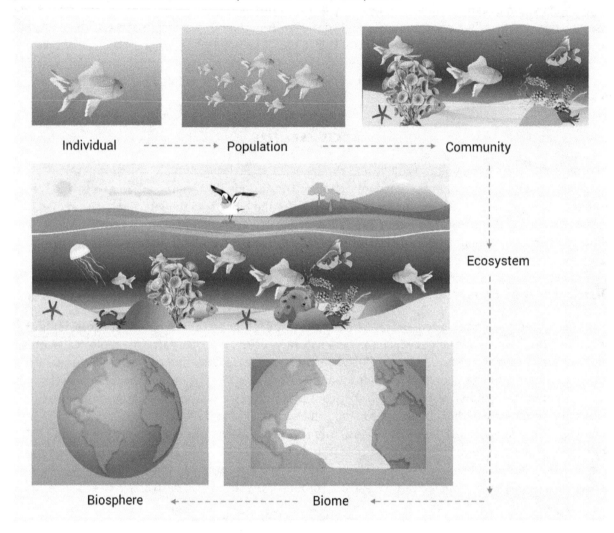

A biosphere is the collection of all of the biomes in the world. A biome is a group of ecosystems with similar properties, such as climate and geography

Conservation Biology

Recycling and conservation are two important tools for protecting the Earth's resources, but as relatively newer practices, they are not without issues. In general, the practice of **recycling** allows for the **conservation** of Earth's resources by reusing manufactured products, which limits the production and use of raw materials. This reduces landfill use, minimizes waste elimination practices that release greenhouse gas emissions, and is often more cost-effective for manufacturers. However, introducing new recycling centers to an area is often costly in the beginning, as it requires constructing and developing the facility and hiring and training workers. Recycling facilities are often dirty, due to the

nature of the items that are recycled, which may have once contained food items, human waste, and other organic materials. These materials quickly rot, may attract vermin, and/or create an overall biological hazard. If the waste from recycled materials is improperly handled, it can cause a pollution problem. Additionally, recycled materials used to create new goods may not be high quality, which can be problematic for the consumer. Finally, recycling is a newer trend that has not yet been adopted on a global scale. Some researchers worry that the amount of recycling that occurs is on a scale that is too small to have a lasting impact, and therefore may be a cost-prohibitive practice.

Recycling is the act of repurposing trash materials that would have otherwise been discarded into a landfill. Discarded materials like paper, cardboard, metal, glass, wood, and plastic can often be used to make new items rather than extracting and utilizing raw materials to do so. This practice reduces waste and creates space in landfills, conserves materials like wood that are slow to renew, reduces methane production (therefore decreasing greenhouse gas production), and reduces energy production. Recyclable products can be sold domestically and globally, creating jobs and goods. Many residential and commercial buildings have municipal or county recycling services that will collect recycled items, similar to garbage collection.

A number of chain grocery stores will take back plastic grocery bags for reuse. Cities, universities, airports, and other large domains are also making the practice of recycling more accessible by setting up single stream recycling containers in public locations, similar to garbage containers. Single stream recycling allows any recyclable material to be placed into the same container for collection. In previous years, consumers often had to separate recyclable goods themselves by material, which limited the number of people who chose to recycle. Products made from recycled materials are often noted as such. For example, cardboard food boxes, fast food napkins, paper towels, and soda cans often display information that they are made from previously recycled products, and from where the product was recycled (a program, a facility, or a consumer).

Environmentally Friendly Consumer Products

Environmentally friendly consumer products are those that are made from recycled components, or are biodegradable. **Biodegradable materials**, even in a landfill, will eventually completely organically decay over time. Environmentally friendly consumer products also refer to daily use household items that are made and utilized in a sustainable manner. For example, some toothpaste and soap manufacturers have removed certain chemicals and non-biodegradable plastics from their products, which consequently keeps plastics out of the soil system and the water supply. Some food companies produce only sustainable items, meaning that the production of the food is not depleting a natural ecosystem. Consumer demand for these types of products has increased greatly, and companies are trying to meet this demand. Eco-friendly companies not only produce environmentally friendly products, but also generally employ environmentally friendly manufacturing practices within their organizations.

Biodiversity

Biodiversity refers to the varied number of species on Earth—ranging from humans, to fish, to plants— and the way ecosystems are built within them. The biodiversity of an area strongly influences its air and soil quality, its energy availability, and how well its community thrives. Natural resources are currently being expended faster rate than they can be replaced, which is resulting in the extinction of species. As all species are interconnected in some way, the loss of an entire species can detrimentally impact the interactions and existence of other species. For example, if a particular animal feeds primarily on a plant species that becomes extinct, the animal species will have to radically change its feeding behaviors or it becomes prone to extinction. Decreasing supplies of water can impact the existence of plant and animal species as well, which, in turn, may affect how and what humans eat and grow. Plants are crucial to providing oxygen and reducing carbon dioxide—a greenhouse gas—on Earth. Additionally, many plants serve as ingredient sources in medicines; loss of plant life affects not only potential food sources but also medicinal sources. Overpopulation is likely the biggest threat to biodiversity, due to the inherent competing needs for land, water, and food production, as well as the risks of excess waste and pollution.

Effects of Human Intervention

Humans often have the biggest impact on ecosystems. Our population size and activities often disrupt the ecosystems of other organisms. Among the most impactful activities of humans on ecosystems is deforestation. Forests are torn down for lumber to create homes for our growing population, as well as to create farmland to feed our growing population. This deforestation destroys the habitats of many organisms and can lead to their death. Many species have become extinct because of deforestation and loss of habitat. Examples of species that are extinct because of deforestation include the Tasmanian tiger, the passenger pigeon, and the Javan tiger. Pandas, orangutans, tigers, and gorillas are among the many species in danger of extinction from deforestation. The larger the human population gets, the more their activities impact ecosystems. What's more, this growing impact occurs at a rate disproportionately higher than the rate of increase of the human population.

Humans can cause change to an ecosystem at an even faster rate than nature. Humans destroy habitats and introduce species that have the potential to be very invasive to new areas. One example of this is the introduction of smallpox to the Western Hemisphere. Smallpox was a disease caused by a virus that had existed in the Eastern Hemisphere since around 10,000 BC. It did not exist in the Western Hemisphere until European exploration in the fifteenth century. The European travelers exposed the native populations of the Western Hemisphere to smallpox. Since this population had no existing immunity, the population was decimated.

The Use and Extraction of Earth's Resources

Extracting resources from the Earth is inherently damaging in its process. **Mining** for minerals and fossil fuels has vast environmental impacts. Surface damage, unnatural erosion, increases in sinkholes, disruption to ecosystems, unnatural animal migration, and pollution are all side effects of mining. **Deforesting** lands to use the land for commercial or residential use or to use the trees for raw materials significantly disrupts ecosystems, contributes to global warming from reduced carbon dioxide consumption, affects water levels, reduces biodiversity, and endangers wildlife. Many rainforests, such as the Amazon rainforest, are believed to have "tipping points" of damage, where the land will be unable to replenish itself and the overall climate will have changed so drastically that it will set off other climate feedback responses. For example, cutting down trees leads to increased atmospheric carbon dioxide in the area, which leads to higher temperatures, which decreases plant water availability, resulting in less vegetation (and the loop continues). **Land reclamation** often focuses on correcting negative impacts to natural resources (i.e., restoring deforested lands by planting indigenous vegetation, replacing sands near beaches that have eroded, and so forth).

Air and Water Pollution

Air and water are vital to human existence. Clean air and potable water greatly impact human health outcomes. As countries develop and become more industrialized, pollution is inevitable. In the United States, the Industrial Revolution, which shifted the economy's focus from agricultural practices to

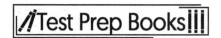

manufacturing, greatly increased air, water, and soil pollution. This occurred because factories burned more coal to operate, leading to increased levels of **smog** (a type of hazy air resulting from the presence of smoke, sulfur oxides, nitrogen oxides, and/or additional hazardous organic compounds) and the presence of **acid rain** (rain that is acidic as a result of air pollution and consequently harms trees, bodies of water, and animals when it falls). Additionally, factories often disposed of waste in the most convenient manner possible—usually by dumping it into bodies of water often used for drinking water. By the middle of the 20th century, both the United States and England had experienced deadly smog events that had resulted in the deaths and sickness of thousands of citizens. These events spurred environmental movements in the 1960s, and the United States passed the **Clean Air Act** in 1970 to combat environmental hazards resulting from air pollution. Currently, many companies focus on the development and implementation of "clean" technologies to manage these issues. The introduction of battery-operated vehicles intended to reduce the country's production of automotive emissions is just one example.

Climate Change and Greenhouse Gases

Greenhouse gases in the Earth's atmosphere include water vapor, carbon dioxide, methane, nitrous oxide, and **chlorofluorocarbons (CFCs)**, which trap heat between the surface of the Earth and the Earth's lowest atmospheric layer, the troposphere. The increase of these gases leads to warming or cooling trends that cause unpredictable or unprecedented meteorological shifts. These shifts can cause natural disasters, affect plant and animal life, and dramatically impact human health. **Water vapor** is a naturally found gas, but as the Earth's temperature rises, the presence of water vapor increases; as water vapor increases, the Earth's temperature rises. This creates a somewhat undesirable loop. **Carbon dioxide** is produced through natural causes, such as volcanic eruptions, but also is greatly affected by human activities, such as burning fossil fuels. A significant increase in the presence of atmospheric carbon dioxide has been noted since the Industrial Revolution; this is important as carbon dioxide is considered the most significant influencer of climate change on Earth. **Methane** is produced primarily from animal and agriculture waste and landfill waste. **Nitrous oxide** is primarily produced from the use of fertilizers and fossil fuels. CFCs are completely synthetic and were previously commonly found in aerosol and other high-pressure containers; however, after being linked to ozone layer depletion, they have been stringently regulated internationally and are now in limited use. Scientists have stated that the climate

shifts recorded since the Industrial Revolution cannot be attributed to natural causes alone, as the patterns do not follow those of climate shifts that took place prior to the Industrial Revolution.

Natural Greenhouse Effect vs. Human Influence

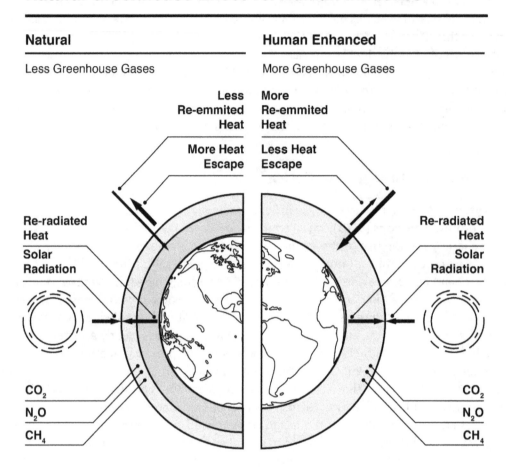

Irrigation

Irrigation refers to a systematic watering method, most pertinent to the agriculture and landscaping industries. It is not the only method of watering used in agricultural practices (some farmers and landscapers do utilize natural rainfall alone), but it is the primary method in which humans can control how to best utilize their most reliable water source for agricultural purposes. Irrigation systems can also be used in conjunction with a community's sewage and drainage system. There are multiple methods of irrigation, although they all have the same goal of supplying crops with a sufficient amount of water. Surface irrigation is the most commonly utilized method of irrigation, functioning by allowing water to freely flow across the desired area and naturally seeping into the soil below. Localized irrigation distributes water through piping and sprays directly onto a plant. An example is a residential sprinkler system that waters an entire lawn through in-ground nozzles, or a drip irrigation system that distributes water using low-flow pressure systems. Subsurface irrigation systems are underground and concentrate water at the root of a plant. Some methods of localized irrigation also combine fertilizer into the watering system.

Reservoirs and Levees

Reservoirs and **levees** are typically utilized as a means of directing and storing water from naturally occurring sources toward and within areas of need. They may be crucial in residential areas that are not close to a naturally existing source of water, such as in desert climates. Resourcing water in this way allows nearby communities to have drinking water, irrigation systems, recreation that centers on water activities, and may also serve as a power-generating source. Levees are used to control flooding of reservoirs, or may be built and used independently in flood-prone areas. Over the decades, both concepts have faced scrutiny from environmental and political experts. Creating a reservoir often utilizes dams, which create a barrier between the natural source of water and the created source of water.

Damming can result in creating unnatural barriers for ecosystems that exist in the naturally occurring water source; sediment build-up in the reservoir, which affects its storage abilities; erosion that lowers water table levels and consequently affects crop output; and human errors such as poor construction. Poorly constructed dams and levees can break and lead to catastrophic flooding for nearby communities, especially those that are downstream from the water source. For example, Hurricane Katrina in 2005 caused enormous destruction to the city of New Orleans, Louisiana, and its surrounding areas as a result of inadequate levee systems that allowed for catastrophic flooding. Additionally, some studies indicate that reservoirs, acting as relatively stagnant water sources, can breed disease. Other studies show that warm-climate reservoirs contribute to excess greenhouse gas production due to the biomass production that occurs over time at the bottom of the reservoirs; this leads to the production and release of methane.

Depletion of Aquifers

Aquifers are naturally occurring sources of extractable freshwater water, normally found in permeable rock. These rocks can be drilled and pumped for water. The availability of water, and if it flows autonomously (without the use of a manmade pump), depends on the type of rock in which it is stored—more porous rocks may allow for more water flow—and seasonal precipitation. The rock also serves as a filter for the water; for example, clay and coal particles can often filter pesticide residue and other hazardous run-off that might taint fresh groundwater. Aquifers are becoming non-potable or depleted primarily from human activity. Residential and commercial use of septic tanks, overuse of fertilizers and pesticides on crops, sustained pumping along ocean coasts, and mining either degrade the quality of the water by allowing hazardous contaminants (such as waste or saltwater) to enter it, or by exposing the water to air and allowing it to evaporate. Depletion occurs when pumping occurs faster than the rocks can replenish their water stores. With exponential population growth, aquifer water is being used at a rate at which it cannot be replenished quickly enough.

Ozone Layer Depletion

Located in the stratosphere, the **ozone layer** protects the Earth from excessive **ultraviolet B (UVB)** ray exposure. The last century has shown signification depletion of the ozone layer, especially over Antarctica; this region is known as the **ozone hole**, missing almost 70% of its ozone layer. Chlorine molecules are especially harmful to ozone molecules. CFCs have been a major contributor to the ozone layer's depletion due to their high concentration of chlorine molecules. Almost all CFC production was a result of industrialization and human activity. In 1996, most CFC production was banned; however, it is expected that atmospheric chlorine levels will remain high for the next couple of decades. Additionally, other effects of climate change may prevent the stratosphere from ever reaching the gas composition that existed before CFCs were utilized. While ozone depletion does not contribute to global warming directly, its impact on human health and disease is significant. The consequent increase in UVB exposure is linked to skin cancer in people, and ecosystem and food source disruption in animals. The effect on plants can lead to plant loss, which can indirectly impact the greenhouse effect, global warming, climate change, and human health.

Waste Disposal and Landfills

Waste disposal is a serious human concern. Waste production has almost doubled in the United States in the last 50 years, with the average household producing over 6,000 pounds of trash per year. Over half of that waste is disposed of in man-made sites in the Earth's ground. Piling (and even burying) trash is an ancient tradition. **Dumps** are open pits of trash, susceptible to rot, stench, and animal infestations. **Landfills** are designed structures intended to create a distinct boundary between the trash and the Earth. This boundary may be made of plastic, or with clay and soil. These structures try to prevent contamination of aquifers and crop soil. As waste breaks down in a landfill, methane is released into the air. Environmental groups and government regulations are pushing lifestyle changes and new technologies to reduce human trash generation. These include repurposing waste, extracting valuable materials from waste, turning waste into a renewable energy source, and advocating green behaviors such as using reusable grocery bags, using fewer plastic goods, and having **compost bins** at home. Many items in landfills could be disposed of in compost bins. These are composed of organic materials that decay quickly, such as food rinds and plant detritus. Once decayed, this material can be used to enrich soil and plant life, limit erosion, and even retain extra groundwater. The average household throws away between 20 to 50 percent of items to landfills that are compostable.

Finally, businesses and landfills in many countries have experienced new regulations aimed to limit waste production. Some governments offer tax breaks to companies that utilize green behaviors and focus on waste reduction.

Practice Questions

1. What happens to the population of a predator if the population of a prey decreases?
   a. It increases rapidly
   b. It decreases
   c. It stays the same
   d. It increases gradually
   e. There is not enough information to answer the question

2. If an ecosystem lost its denitrifying bacteria, where and in what form would nitrogen accumulate?
   a. In the soil as nitrates
   b. In the air as N_2
   c. In the soil as ammonium
   d. In the air as nitrates
   e. In the soil as N_2

3. An ecosystem that normally has moderate summers with high rainfall is experiencing a heat wave and a drought. How does this affect the rate of photosynthesis of the producers in this ecosystem?
   a. The decrease in transpiration from the high heat and the drop in rainfall decreases the number of chloroplasts, so photosynthesis rates decrease.
   b. The increase in transpiration from the high heat and the drop in rainfall results in less water. Since photosynthesis creates water, the rate increases to meet increased water demands.
   c. Water availability has no effect on photosynthesis.
   d. Increased temperature increases the number of mitochondria, so photosynthesis rates increase.
   e. The increase in transpiration from the high heat and the drop in rainfall results in less water available for photosynthesis. The rate decreases.

4. What is the role of an allosteric activator?
   a. To bind to the active site of an enzyme and block the binding of a substrate.
   b. To bind to the active site of an enzyme and allow the binding of a substrate.
   c. To bind to an unrelated site of an enzyme to block the binding of a substrate.
   d. To bind to an unrelated site of an enzyme to allow the binding of a substrate.
   e. To circulate freely and increase the rate of enzyme production.

5. A new species is introduced into an ecosystem. This species is a parasite to an existing species in the ecosystem, the host. What will the immediate effects be on the population size of the parasite and the host?
   a. Parasite increases, host decreases
   b. Parasite increases, host increases
   c. Parasite decreases, host increases
   d. Parasite decreases, host decreases
   e. The host will adapt to become resistant to the parasite.

6. A farmer grows all of his tomato plants by vegetative propagation. He finds that one clone produces tomatoes that sell much better than any other clone. He then uses this clone to plant his entire field. Two years later a fungus wipes out his entire crop. What could the farmer have done to prevent this?

I. Plant tomatoes that sell poorly. They are more resistant to fungus.
II. Plant a variety of tomatoes. Genetic variation would have left some of the crop less susceptible to the fungus.
III. Plant a variety of crops. Plants other than tomatoes might not be affected by the fungus.

a. Choice I only
b. Choice II only
c. Choice III only
d. Choice I or III
e. Choice II or III

7. An ecosystem experiences a loss of one species due to hunting. While the overall population size of this species was small, the loss of this species is devastating to the ecosystem. What kind of species was this?
a. Producer
b. Primary Consumer
c. Parasite
d. Keystone
e. Secondary consumer

8. To answer the following question, refer to the diagram below:

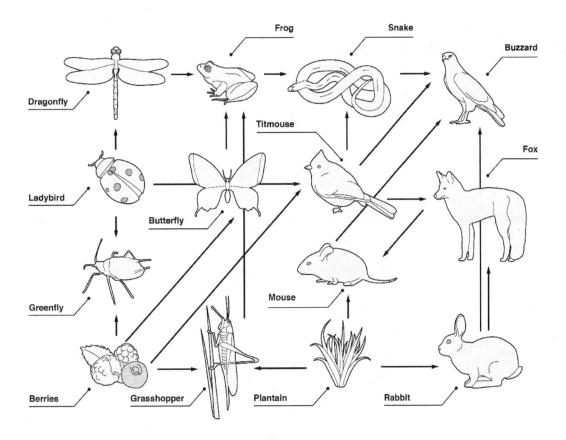

Assume that the snake population has been wiped out by a disease that is only transmutable between snakes. Which of the following would most likely be true as a result?
   a. The dragonfly population would decrease.
   b. The grasshopper population would increase.
   c. The fox population would decrease.
   d. The buzzard population would increase.
   e. The titmouse population would decrease.

9. Amphibian development is different than mammalian development in that it involves a larval stage and metamorphosis. Unlike the caterpillar/butterfly metamorphosis cycle, the stages of larval development in the tadpole/frog conversion are readily observable. Identify the most likely mechanism for the exchange of the tail for legs in frog development.

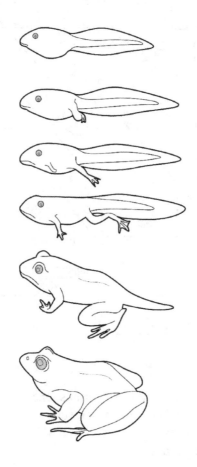

a. The tails are destroyed via necrosis, causing them to grow legs in order to replace the tail.
b. While the organs are developing, homeotic gene factors induce formation of the notochord and primitive brain, which signals the posterior region to die.
c. Tail cells undergo apoptosis, which slowly degrades tail tissue while cell determinants initiate leg development.
d. A feedback loop in the frog evaluates environmental conditions and sends hormones to initiate cell death in the tail when water is scarce.
e. Tails require more metabolic activity so as the frog matures and the body size changes, insufficient energy is available to support a tail, so it dies.

10. The homology between the circulatory systems of different vertebrates is illustrated in the image below. Which statement best explains the similarities and differences in the physiology between the different organisms?

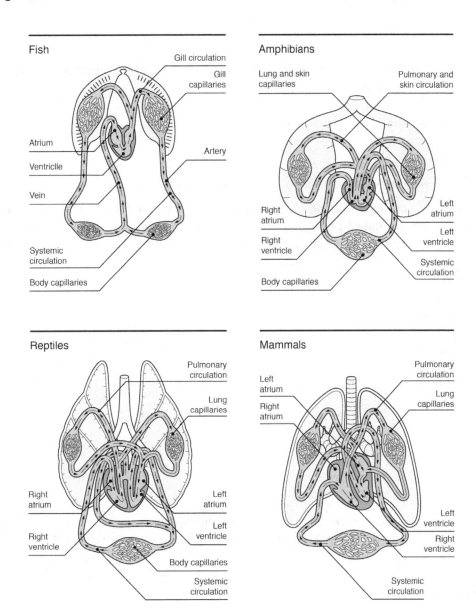

a. The large conservation between amphibians and reptiles suggests that reptiles are the direct descendants of amphibians.

b. Fish are the only group pictured that has gill capillaries, suggesting that there is no common ancestor between fish and the terrestrial groups.

c. The diagrams suggest that reptiles and mammals are more closely related than reptiles and amphibians.

d. The significant difference in the circulatory system between mammals and fish implies that every mammalian organ is more advanced.

e. The significant conservation between fish and amphibians suggests that amphibians are the direct descendants of fish.

11. Which period in history dramatically increased air, water, and soil pollution?
   a. The Paleolithic Era
   b. The Big Bang Era
   c. The Industrial Revolution
   d. The Medieval Ages
   e. The Renaissance

12. Which of the following is a man-made compound that is no longer approved for use due to its major detrimental effect on the ozone layer?
   a. Teflon
   b. Chlorofluorocarbons
   c. Carbon trioxide
   d. Fluorocarbon phosphates
   e. Zinc oxide

13. Angie, a second-year college student, is visiting the Gulf Coast of Florida with an environmental conservation group for alternative spring break. The group's efforts focus on transporting truckloads of sand from various locations to a particular beach and creating dunes along some of the beach grasses. What conservation practice is this an example of?
   a. Deforestation
   b. Grassroots movement
   c. Bioremediation
   d. Recycling
   e. Reclamation

14. Which greenhouse gas is a common byproduct of landfills and concentrated animal feeding operations?
   a. Carbon
   b. Corn fumes
   c. Nitrogen
   d. Methane
   e. Sulfur

15. Which of the following substances contributed to a lethal fog that spread over areas of England the United States in the mid-1900s, killing thousands and making many more critically ill?
   a. CFCs
   b. Smog
   c. Alkaline rain
   d. Cow manure
   e. Arsenic

16. What percentage of the average household's trash could actually be composted, therefore reducing landfill usage?
   a. 10% to 20%
   b. 50% to 60%
   c. 5% to 20%
   d. 20% to 50%
   e. 70% or more

Answer Explanations

1. B: When the population of a prey decreases, the population of the predator will also decrease as competition increases between the individuals in the predator population and as the prey resource becomes scarce.

2. A: Denitrifying bacteria live in the soil and convert nitrates to N_2. If they did not exist, nitrates would accumulate in the soil.

3. E: Water is essential for photosynthesis. Increasing temperatures increase transpiration and drought conditions result in less water available for photosynthesis. The rate of photosynthesis will decrease.

4. D: Allosteric activators bind to an allosteric site, a site other than the active site, of an enzyme and cause a conformational change that allows the substrate to bind to the active site of the enzyme.

5. A: A parasite benefits from the relationship with the host, while the host suffers a fitness cost. Therefore, the population of the parasite will increase while the population of the host will decrease. Choice *E* is incorrect because the question asks for the immediate effect; while perhaps over time some hosts may adapt to be resistant to the parasite, this would not happen immediately. Adaptation takes time.

6. E: Genetic variety in a species allows them to be more resistant to stresses. Having genetic diversity increases resilience. Growing multiple strains of tomatoes or multiple types of crops could protect the farm.

7. D: Keystone species are any species whose role in the ecosystem is disproportionate to the size of the population. The loss of a keystone species will devastate an ecosystem.

8. A: If the snake population were to disappear, the organisms that it hunted would increase, and the food that those organisms ate would decrease. Without the snake, that means there are more frogs, which in turn means there are more frogs to eat more dragonflies, so the dragonfly population would decrease. Choice *B* is incorrect, since an increase in frogs means more frogs would eat more grasshoppers, so grasshoppers would decrease as opposed to increase. Choices *C* and *E* are false because fewer snakes means more titmice (not fewer), which provide more food for the foxes and results in an increase in the fox population, not a decrease. Choice *D* is incorrect because fewer snakes means less food for the buzzards to eat, resulting in a decrease in population.

9. C: Tadpole tail degradation is a classic example of apoptosis, or programmed cell death, which is an important process in morphogenesis of organisms. Apoptosis explains why some animals have webbed feet and some don't. Choice *A* is incorrect because necrosis is cell death induced by injury, and leg formation in tadpoles is not caused by injury. Choice *B* is wrong because apoptosis is not a nervous system-controlled event; it is either activated by an individual cell's internal signals or neighboring cell signals. Choice *D* is incorrect because apoptosis is not controlled by hormones. Choice *E* is nonsensical because the frog develops legs, which is a highly metabolically-demanding activity.

10. C: The conservation of the four-chambered heart provides the justification for concluding that the greater relatedness is between mammals and reptiles. As this is somewhat subjective due to qualitative interpretation, it is still the best choice. Choices *A* and *E* are incorrect because organisms are not necessarily direct descendants of more primitive versions of existing organisms; they could have branched from along different points in the phylogenetic tree. Choice *B* is incorrect because all organisms share a common ancestor. Choice *D* is incorrect because it is impossible to extrapolate broad conclusions based on qualitative data regarding one body system. Fish do in fact have very complicated organs such as gills that are highly differentiated and complex.

11. C: The Industrial Revolution switched the economical focus for most of the world from agriculture to manufacturing. This period produced factories and many machines, which required the combustion of coal and other fuel sources. As a new industry, the lack of regulation did not combat the air pollution from these factories, nor were there rules on where to dump waste. While some pollution likely did occur in the other periods listed, the period of the Industrial Revolution, from approximately the mid-1760s to the early 1800s, caused a dramatic spike.

12. B: Chlorofluorocarbons (CFCs) were man-made compounds of chlorine, fluorine, and carbon used mainly in aerosol cans and refrigerants. Their use single-handedly caused significant depletion to the ozone layer, at a rate of about 20% globally and 70% over Antarctica. CFCs have mostly been banned from production, but it is unclear whether the ozone layer will ever reach pre-CFC usage levels. Teflon and zinc oxide do not directly contribute to ozone layer depletion and the other options listed are not real compounds.

13. E: Reclamation refers to making an area of defiled land usable by returning it to its natural state. In this instance, the spring break group is helping the shoreline by preventing further erosion from destroying the beach and its vegetation. Deforestation refers to removing vegetation. Grassroots movements typically refer to local, organized political movements. Bioremediation uses bacteria or other microbes to decontaminate soil and groundwater. Recycling typically refers to repurposing a material that has already been consumed and making into another usable item.

14. D: Landfills and animal waste are large contributors to methane. They do not cause significant amounts of the other greenhouse gases listed. Corn fumes are not a greenhouse gas.

15. B: Smog produced from organic, toxic compounds from factories so heavily blanketed areas in England and the United States that many suffocated or became violently ill. This eventually led to air pollution initiatives. The other items listed did not play a role.

16. D: A large percentage of organic household items could be composted rather than be placed in landfills. Residential composting bins are becoming more popular in modern society, but could be advocated further as an easy way to reduce household waste.

Genetics

Meiosis

Meiosis is a type of cell division in which the daughter cells have half as many sets of chromosomes as the parent cell. In addition, one parent cell produces four daughter cells. Meiosis has the same phases as mitosis, except that they occur twice—once in meiosis I and once in meiosis II. The diploid parent has two sets of chromosomes, set A and set B. During meiosis I, each chromosome set duplicates, producing a second set of A chromosomes and a second set of B chromosomes, and the cell splits into two. Each cell contains two sets of chromosomes. Next, during meiosis II, the two intermediate daughter cells divide again, producing four total haploid cells that each contain one set of chromosomes. Two of the haploid cells each contain one chromosome of set A and the other two cells each contain one chromosome of set B.

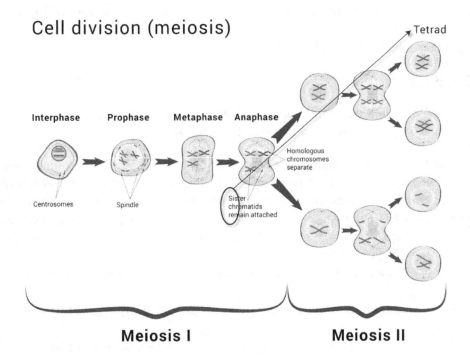

Meiosis is similar to mitosis because it involves cellular division. However, while mitosis involves the division of somatic (body) cells, meiosis is specifically the production of gametes (eggs and sperm). In mitosis, one parent cell splits once into two genetically identical and diploid daughter cells, while in meiosis, one germ cell splits twice into four genetically different, haploid daughter cells. The two divisions of meiosis (**meiosis I** and **meiosis II**) are critical because, when a sperm fertilizes an egg to create the first cell of a new organism (**zygote**), the zygote must have two sets of chromosomes—not four—to be viable.

The phases of meiosis I and II are nearly the same as the phases of mitosis, except for a few key differences. In mitosis, homologous chromosomes line up single file in metaphase. On the other hand, **tetrads** are paired homologous chromosomes, or homologs, which are lined up in metaphase. The tetrad is held together by the **synaptonemal complex**, which connects pairs of homologous chromosomes and is disassembled at the end of prophase I. In prophase I, tetrads additionally go through a process called **crossing over/recombination** where they exchange DNA. A **chiasmata** is where the crossing over occurs (there are 1-3 crossing over events per tetrad). Crossing over makes gametes unique because genetic recombination events are random and unpredictable. In anaphase I, the homologous chromosomes separate, moving pairs of sister chromatids (replicated chromosomes) to each side of the cell.

After meiosis I, the homologous chromosomes have been separated, and the two daughter cells are **haploid**. There is no interphase between meiosis I and meiosis II, so the DNA is not replicated. In anaphase II of meiosis, sister chromatids separate and, by the end, there are four unique, haploid daughter cells.

Although homologous chromosomes code for the same genes, there is diversity. This is because each gene has two or more **alleles**. Alleles are different forms of genes, and the **phenotype** is dependent on the **genotype**.

A process called crossing over occurs, which makes the daughter cells genetically different. If chromosomes didn't cross over and rearrange genes, siblings could be identical clones. There would be no genetic variation, which is a critical factor in the evolution of organisms.

Stages of Meiosis

In summary, nearly the same stages occur in meiosis as in mitosis, but twice, with some important differences. The five stages of meiosis I occur in order as follows:

- *Prophase I*: The nuclear envelope breaks up and DNA condenses into chromosomes. Replicated chromosomes group up to form homologous chromosome pairs, called tetrads, connected by the synaptonemal complex. Crossing over occurs, resulting in unique sister chromatids.
- *Prometaphase I*: The nuclear membrane breaks apart, the synaptonemal complexes between homolog pairs are dismantled, tetrads develop kinetochores at their centromeres, and the mitotic spindle begins to form.
- *Metaphase I*: Tetrads are lined up along the metaphase plate and connect to the mitotic spindle.
- *Anaphase I*: Pairs of homologous chromosomes separate, giving a pair of sister chromatids to each side.
- *Telophase I*: Paired sister chromatid are located on each side of the cell prior to cytokinesis.

After meiosis I, cytokinesis occurs, splitting the cell into two haploid cells, with each containing two sister chromatids. No DNA replication occurs between these phases.

- *Prophase II & Prometaphase II*: Largely the same as mitosis, the (non-identical) sister chromatids, which are still connected in pairs from meiosis, form kinetochores at their centromeres and the mitotic spindle reforms.
- *Metaphase II*: The mitotic spindle attaches to the chromosomes and they line up once again at the metaphase plate.
- *Anaphase II*: Sister chromatids are separated and move apart to each side of the cell.
- *Telophase II*: Each half now has one of each chromatid. The nuclear envelope reforms, the mitotic spindle disassembles, and the chromosomes decondense before the final cytokinesis.

The daughter cells after meiosis II occurs are haploid, each having one copy of each chromosome. When two gametes are fused in sexual reproduction, this creates a viable diploid cell called a zygote.

Mendelian Genetics

A monk named Gregor Mendel is referred to as the father of genetics. He was responsible for coming up with one of the first models of inheritance in the 1860s. His model included three laws to determine which traits are inherited. These laws still apply today, even after genetics has been studied much more in depth.

Chromosomal Basis of Inheritance

Tenets of Mendelian genetics:

- *The law of dominance*: Dominant alleles trump recessive alleles in the phenotype (the exceptions are non-Mendelian traits)

- *The law of segregation*: Alleles for each trait are separated into gametes. One allele comes from each parent, giving the offspring two copies of each allele.

- *The law of independent assortment*: Tetrads line up in metaphase I independently of other chromosomes. Each of the 23 homologues has a 50/50 chance of being on either side.

In simple Mendelian genetics, an individual can have three different genotypes. This is shown in the table below regarding the trait of flower color:

Genotype	Referred to as	Corresponding Phenotype
PP	Homozygous dominant	Purple flowers
Pp	Heterozygous	Purple flowers
pp	Homozygous recessive	White flowers

	Height	Seed Shape	Flower Color
Dominant	Tall	Round	White
Recessive Trait	Short	Wrinkled	Violet

When Mendel crossed true breeds in the P generation (as shown in the Punnett square below), he noted that all the F_1 offspring had the dominant phenotype. When he crossed the F_1 offspring (all heterozygotes), the F_2 generation consistently showed a 3:1 phenotypic ratio of purple to white flowers. These results demonstrate the first law of Mendelian genetics: the law of dominance.

Parent (P) generation: True breed cross: *PP* x *pp*

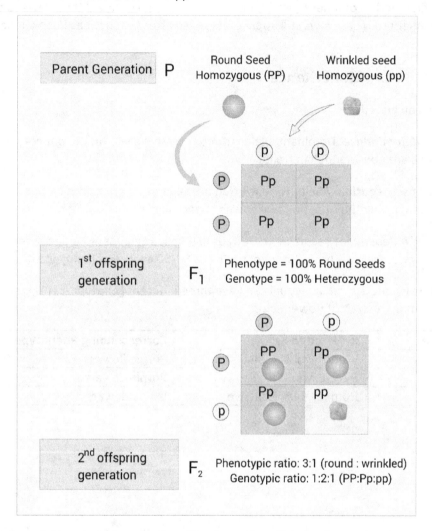

A useful application of this rule is the practice of **backcrossing**. This is when a dominant expressing organism of an unknown genotype is crossed with a recessive expressing organism. Offspring phenotype can be used to determine if the parent is homozygous or heterozygous.

Dihybrid crosses involve two traits and illustrate Mendel's law of segregation: alleles separate in meiosis.

Dihybrid crosses example: Cross *PpRr* x *PpRr*

- Purple flowers = *P* (dominant)
- White flowers = *p* (recessive)
- Round seeds = *R* (dominant)
- Wrinkled seeds = *r* (recessive)

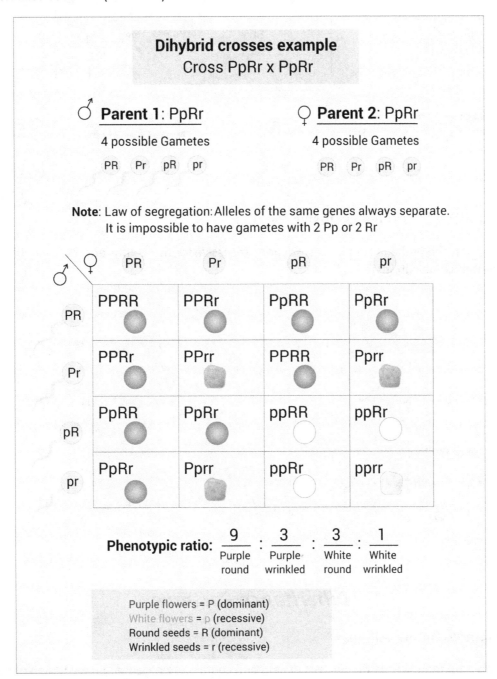

Probability can be determined by using the **law of multiplication** when all factors are present. The **law of addition** can be used to determine probability when one factor OR others may be present. Another dominant phenotype is yellow seed (*Y*) over green seed (*y*), and tall stems (*T*) are dominant over dwarf stems (*t*).

Question: What is the probability of having offspring with the genotype *PPrrYytt* if the parent cross is *PprrYyTt* x *PpRryyTt*?

Solution:

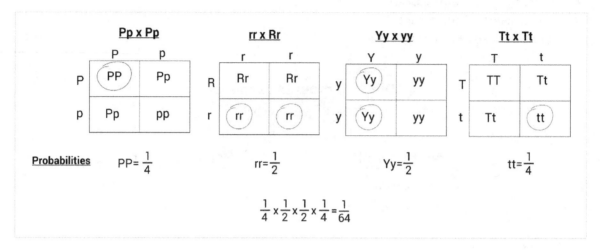

Law of Addition (Mutually Exclusive Results)

Question: What is the probability of having offspring recessive for three different traits (could contain several different combinations)?

Solution:

- $\frac{Pp}{pp}$ rryytt $= \frac{3}{4} \times \frac{1}{2} \times \frac{1}{2} \times \frac{1}{4} = \frac{3}{64}$

- pp$\frac{RR}{Rr}$ yytt $= \frac{1}{4} \times \frac{1}{2} \times \frac{1}{2} \times \frac{1}{4} = \frac{1}{64}$

- pprr $\frac{Yy}{yy}$ tt $= \frac{1}{4} \times \frac{1}{2} \times \frac{1}{2} \times \frac{1}{4} = \frac{1}{64}$

- pprryy $\frac{TT}{Tt} = \frac{1}{4} \times \frac{1}{2} \times \frac{1}{2} \times \frac{3}{4} = \frac{3}{64}$

ADD • mutually exclusive

$\frac{3}{64} + \frac{1}{64} + \frac{1}{64} + \frac{3}{64} = \frac{8}{64}$ $\frac{8}{64} = \frac{1}{8}$

Inheritance Patterns

Chromosomes, Genes, Alleles

Chromosomes are found inside the nucleus of cells and contain the hereditary information of the cell in the form of **genes**. Each gene has a specific sequence of DNA that eventually encodes proteins and results in inherited traits. **Alleles** are variations of a specific gene that occur at the same location on the chromosome. For example, blue and brown are two different alleles of the gene that encodes for eye color.

Dominant and Recessive Traits

In genetics, **dominant alleles** are mostly noted in italic, capital letters (*A*) and **recessive alleles** are mostly noted in italic, lower case letters (*a*). There are three possible combinations of alleles among dominant and recessive alleles: *AA*, *Aa* (known as a heterozygote), and *aa*. Dominant traits are phenotypes that appear when at least one dominant allele is present in the gene. Dominant alleles are considered to have stronger phenotypes and, when mixed with recessive alleles, will mask the recessive trait. The recessive trait would only appear as the phenotype when the allele combination is "*aa*" because a dominant allele is not present to mask it.

A gene can be pinpointed to a **locus**, or a particular position, on DNA. It is estimated that humans have approximately 20,000 to 25,000 genes. For any particular gene, a human inherits one copy from each parent for a total of two. Genotype refers to the genetic makeup of an individual within a species. Phenotype refers to the visible characteristics and observable behavior of an individual within a species.

Genotypes are written with pairs of letters that represent alleles. Alleles are different versions of the same gene, and, in simple systems, each gene has one dominant allele and one recessive allele. The letter of the dominant trait is capitalized, while the letter of the recessive trait is not capitalized. An individual can be homozygous dominant, homozygous recessive, or heterozygous for a particular gene. **Homozygous** means that the individual inherits two alleles of the same type while **heterozygous** means inheriting one dominant allele and one recessive allele.

If an individual has homozygous dominant alleles or heterozygous alleles, the dominant allele is expressed. If an individual has homozygous recessive alleles, the recessive allele is expressed. For example, a species of bird develops either white or black feathers. The white feathers are the dominant allele, or trait (*A*), while the black feathers are the recessive allele (*a*). Homozygous dominant (*AA*) and heterozygous (*Aa*) birds will develop white feathers. Homozygous recessive (*aa*) birds will develop black feathers.

Genotype (genetic makeup)	Phenotype (observable traits)
AA	white feathers
Aa	white feathers
aa	black feathers

Influence of Phenotype on Genotype

The genetic material (DNA) inherited from an individual's parents determines genotype. Natural selection leads to adaptations within a species, which affects the phenotype. Over time, individuals within a species with the most advantageous phenotypes will survive and reproduce. As result of reproduction, the subsequent generation of phenotypes receives the fittest genotype. Eventually, the individuals within a species with genetic fitness flourish and those without it are erased from the environment. As explained above, this is also referred to as the concept of "survival of the fittest". When this process is duplicated over numerous generations, the outcome is offspring with a level of genetic fitness that meets or exceeds that of their parents.

Genetic Crosses

Genetic crosses are the possible combinations of alleles, and can be represented using Punnett squares. A monohybrid cross refers to a cross involving only one trait. Typically, the ratio is 3:1 (*DD, Dd, Dd, dd*), which is the ratio of dominant gene manifestation to recessive gene manifestation. This ratio occurs when both parents have a pair of dominant and recessive genes. If one parent has a pair of dominant genes (*DD*) and the other has a pair of recessive (*dd*) genes, the recessive trait cannot be expressed in the next generation because the resulting crosses all have the *Dd* genotype. A dihybrid cross refers to one involving more than one trait, which means more combinations are possible. The ratio of genotypes for a dihybrid cross is 9:3:3:1 when the traits are not linked. The ratio for incomplete dominance is 1:2:1, which corresponds to dominant, mixed, and recessive phenotypes.

Co-Dominance and Multiple Alleles

The simple dominant/recessive model for genetics does not work for many genes.

For example, blood type is a trait that has multiple alleles: I^A, I^B, and i. I^A and I^B are **co-dominant** so neither is "stronger" than the other, and i is recessive to both. In the event that both co-dominant alleles are present in a genotype, both phenotypes will be present.

Genotype	Phenotype	Blood Donation Facts
$I^A I^A$, $I^A i$	A blood (A antigens and B antibodies)	People with A blood can't receive blood from AB or B due to antibody recognition and attack of B antigen.
$I^A I^B$	AB blood (A and B antigens but no antibodies)	Universal receiver because it contains no antibodies against A or B antigens.
$I^B I^B$, $I^B i$	B blood (B antigens and A antibodies)	People with B blood can't receive blood from AB or A due to antibody recognition and attack of A antigen.
ii	O blood (A and B antibodies)	Can only receive from other O blood (universal donor)

Blood type demonstrates the concept of co-dominance as well as multiple alleles. Below are some blood type crosses and probabilities.

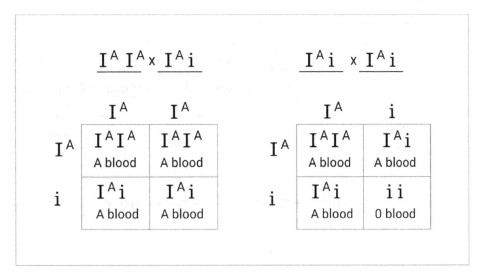

Incomplete dominance occurs when the phenotype is a blending of the two alleles instead of one being dominant over the other. For example, if black and white feathers are co-dominant in birds, heterozygous offspring will have black and white speckles. However, if black and white feathers have an incomplete dominant pattern, heterozygotes will appear grey.

Sex-Linked Genetics

Normal Human Karyotype

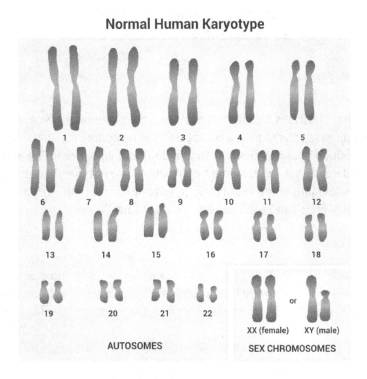

Chromosome 23 is not always homologous because it determines gender, and an individual is either XX (female) or XY (male).

The X chromosome is much larger than the Y chromosome and carries more genes, including the color-blind recessive allele. The possible genotypes, phenotypes, and inheritance patterns are shown below for the color-blind trait:

Genotype	Gender	Phenotype
$X^C X^C$	Female	Normal vision
$X^C X^c$	Female	Normal vision *She carries the allele and can pass it on to her children
$X^c X^c$	Female	Color blind
$X^C Y$	Male	Normal vision
$X^c Y$	Male	Color blind

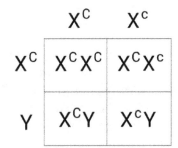

Normal vision male and female carrier

	X^C	X^c
X^C	$X^C X^C$	$X^C X^c$
Y	$X^C Y$	$X^c Y$

- 25% chance of a color - blind child.
- 50% chance of son being color - blind.

Pleiotropy

Pleiotropy is the term that describes a gene that has more than one phenotype. For example, cystic fibrosis is a debilitating disease caused by the mutation of gene coding for a channel protein, which results in mucus overproduction and symptoms that include difficulty breathing and digestive problems. However, the majority of genes have some effect on multiple phenotypes, rather than a single phenotype, which is most evident when examining genetic disorders with varied symptoms. Any gene with multiple phenotypic effects can be described as pleiotropic.

Polygenic Inheritance

A phenotype dependent on more than one gene is called **polygenic inheritance**. It is evident when there is a distribution of phenotypes over a wide range, such as skin color. This is in contrast to **monogenic inheritance** when a phenotype is dependent on only one gene. With polygenic inheritance, several

genes have an additive effect on the same trait, creating a broad gradient of phenotypes depending on the number of same alleles for that trait, which is the result of incomplete dominance.

Epistasis

Epistasis occurs when one gene alters the phenotype of a different gene. For example, in certain mice fur color is determined by the presence of an *Agouti* allele. It codes for brown fur, which is dominant over black fur. Color is also determined by a second allele that codes for pigmentation. The dominant allele *C* means that fur pigment will be made while the recessive allele *c* codes for **albinism** (absence of pigment). The gene for pigment deposition can silence the gene for fur color. Possible genotypes/phenotypes are:

- *AACC, AaCc* = Brown fur
- *aaCC, aaCc* = Black fur
- *AAcc, Aacc, aacc* = White fur

Consider the following possible combinations resulting from epistasis of two brown mice, each with the genotype *AaCc*:

	AC	*aC*	*Ac*	*Ac*
AC	*AACC* brown	*AaCC* brown	*AACc* brown	*AaCc* brown
aC	*AaCC* brown	*aaCC* black	*AaCc* brown	*aaCc* black
Ac	*AACc* brown	*AaCc* brown	*AAcc* white	*Aacc* white
ac	*AaCc* brown	*aaCc* black	*Aacc* white	*Aacc* white

The resultant phenotypic ratios are as follows: white (4/16), black (3/16), brown (9/16).

Pedigrees

Pedigrees show family ancestry and can be used to track genetic diseases through phenotypes. Circles represent females and squares represent males. Shaded shapes represent affected individuals. The genotypes of individuals can be deduced from given phenotypes. For example, below is a pedigree tracing color blindness. Individual I-1 must be X^cY because he is affected. Individual II-6 must be X^cX^c

because, in order for her son to be affected, she had to pass along the color-blind allele she received from her father.

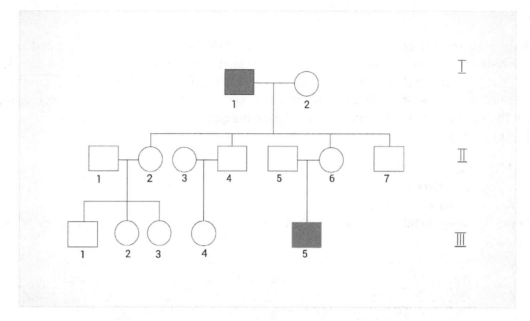

There are two types of **autosomal** or non-sex chromosomes (chromosomes 1-22): dominant and recessive.

The following pedigree tracing Huntington's disease is an example of an autosomal dominant disease. Every individual that has the dominant allele is affected.

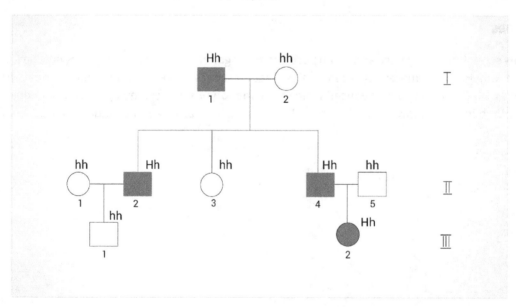

This pedigree tracing albinism is an example of an autosomal recessive disease.

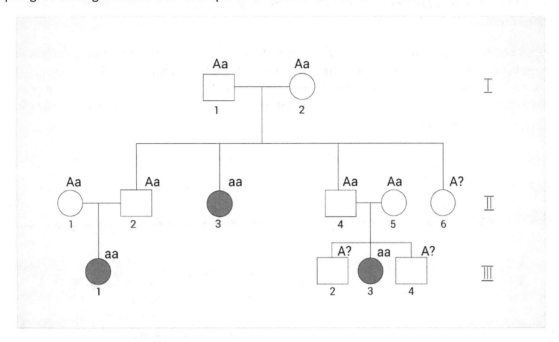

Analysis of this pedigree shows that:

- II-3, III-1, and III-3 have to be aa because they are affected.
- Any couple with children who are affected must both be heterozygotes.
- This particular pedigree can't be a sex-linked, recessive pedigree. In order to be one, an affected female must have an affected father so she can inherit his recessive allele.

Karyotypes

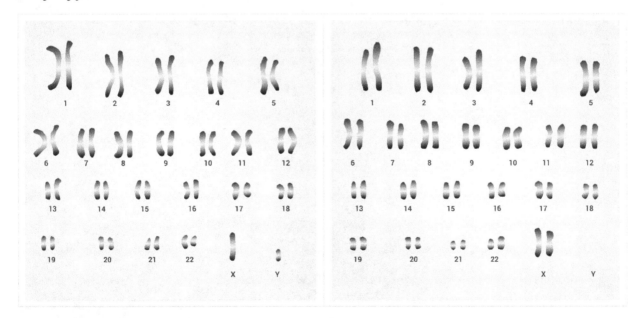

Karyotypes show a picture of an individual's 23 chromosomes and illustrate the diploid nature of our genome. Although females have two X chromosomes, only one X will be active in each cell and the other will form an inactive **Barr body**. The inactive X chromosome is random. Some cells will have one inactive X chromosome while others have the second X chromosome inactive, resulting in an individual consisting of a mosaic of cells.

Karyotypes not only show gender, they can also illustrate occasional mistakes that occur in meiosis called **nondisjunction**. Nondisjunction results in improper separation of tetrads and chromatids. This results in one cell having an extra copy of a chromosome (**trisomy**) and another cell missing a copy of the homologous chromosome (**monosomy**). Fertilization with gametes affected by nondisjunction is often fatal, but some are viable.

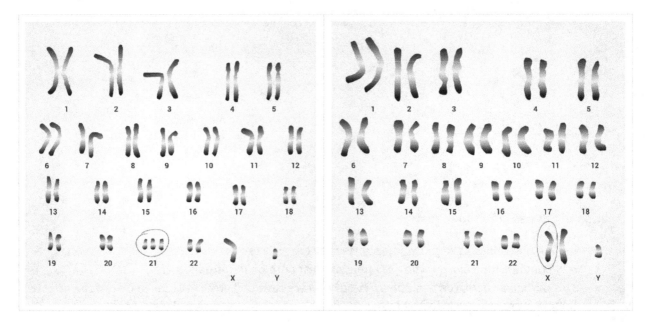

Trisomy 21: Down Syndrome   Trisomy 23: Klinefelter Syndrome

Different diseases and their inheritance patterns are listed below:

Disease	Phenotype	Cause/Pattern
Color Blindness	Red/green vision deficiency	Sex-linked recessive
Hemophilia	Blood clotting disorder	Sex-linked recessive
Muscular Dystrophy	Weak muscles and poor muscle coordination	Sex-linked recessive
Huntington's	Nervous system degeneration that has a late onset (middle age)	Autosomal dominant
Achondroplasia	Dwarfism	Autosomal dominant
Cystic Fibrosis	Excessive mucus production	Autosomal recessive
PKU	Unable to digest phenylalanine	Autosomal recessive
Tay-Sachs Disease	Intellectually disabled due to inability to metabolize lipids, death in infancy	Autosomal recessive
Sickle Cell Anemia	Red blood cell shaped like a sickle instead of a circle	Autosomal co-dominant

Disorders caused by nondisjunction and chromosomal deletions:

Down Syndrome	Intellectually disabled	Trisomy 21
Klinefelter Syndrome	Sterile males	Trisomy 23 (XXY)
Turner Syndrome	Sterile females	Monosomy 21 (X)

Molecular Genetics

Structure and Function of DNA

Deoxyribonucleic acid (DNA) is life's instruction manual. It is double stranded and directional, meaning it can only be transcribed and replicated from the **3' end** to the **5' end**.

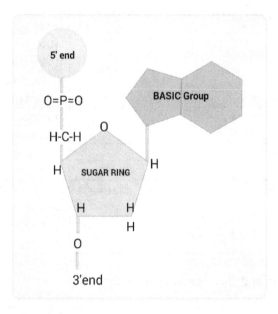

DNA as a Monomer	DNA as a Polymer
A nucleotide is composed of a five-carbon sugar with a Phosphate Group off of the 5th carbon and a Nitrogen Base off of the 1st carbon. DNA and RNA are different because DNA contains deoxyribose sugar while RNA contains ribose sugar. Also, the nitrogen base thymine in DNA is replaced by uracil in RNA.	The two strands are antiparallel, meaning they are read in opposite directions. The bases guanine and cytosine are complementary and held together by three hydrogen bonds, and the bases adenine and thymine are complementary and held together by two hydrogen bonds. Weak hydrogen bonding between bases allows DNA to be opened easily for transcription and replication.

Nitrogen bases come in two varieties:

- **One-ringed pyrimidines**: cytosine and thymine (DNA)/uracil (RNA)
- **Two-ringed purines**: adenine and guanine

Purines and Pyrimidines

The five bases in DNA and RNA can be categorized as either pyrimidine or purine according to their structure. The pyrimidine bases include cytosine, thymine, and uracil. They are six-sided and have a single ring shape. The purine bases are adenine and guanine, which consist of two attached rings. One ring has five sides and the other has six. When combined with a sugar, any of the five bases become nucleosides. Nucleosides formed from purine bases end in "osine" and those formed from pyrimidine

bases end in "idine." Adenosine and thymidine are examples of nucleosides. Bases are the most basic components, followed by nucleosides, nucleotides, and then DNA or RNA.

RNA

RNA is short for ribonucleic acid, which is a type of molecule that consists of a long chain (polymer) of nucleotide units. RNA and DNA differ in terms of structure and function. RNA has a different sugar than DNA. It has ribose rather than deoxyribose sugar. The RNA nitrogenous bases are adenine (A), guanine (G), cytosine (C), and uracil (U). Uracil is found only in RNA and thymine is found only in DNA. RNA consists of a single strand and DNA has two strands. If straightened out, DNA has two side rails. RNA only has one "backbone," or strand of sugar and phosphate group components. RNA uses the fully hydroxylated sugar pentose, which includes an extra oxygen compared to deoxyribose, which is the sugar used by DNA. RNA supports the functions carried out by DNA. It aids in gene expression, replication, and transportation.

RNA acts as a helper to DNA and carries out a number of other functions. Types of RNA include ribosomal RNA (rRNA), transfer RNA (tRNA), and messenger RNA (mRNA). Viruses can use RNA to carry their genetic material to DNA. Ribosomal RNA is not believed to have changed much over time. For this reason, it can be used to study relationships in organisms. Messenger RNA carries a copy of a strand of DNA and transports it from the nucleus to the cytoplasm. DNA unwinds itself and serves as a template while RNA is being assembled. The DNA molecules are copied to RNA. Translation is the process whereby ribosomes use transcribed RNA to put together the needed protein. Transfer RNA is a molecule that helps in the translation process, and is found in the cytoplasm. Ribosomal RNA is in the ribosomes.

Transcription and Translation

Transcription is the process by which a segment of DNA is copied onto a working blueprint called RNA. Each gene has a special region called a promoter that guides the beginning of the transcription process. RNA polymerase unwinds the DNA at the promoter of the needed gene. After the DNA is unwound, one strand or template is copied by the RNA polymerase by adding the complementary nucleotides, G with C, C with G, T with A, and A with U. Then, the sugar phosphate backbone forms with the aid of RNA polymerase. Finally, the hydrogen bonds joining the strands of DNA and RNA together are broken. This forms a single strand of messenger RNA or mRNA.

Codons are groups of three nucleotides on the messenger RNA, and can be visualized as three rungs of a ladder. A codon has the code for a single amino acid. There are 64 codons but 20 amino acids. More than one combination, or triplet, can be used to synthesize the necessary amino acids. For example, AAA (adenine-adenine-adenine) or AAG (adenine-adenine-guanine) can serve as codons for lysine. These groups of three occur in strings, and might be thought of as frames. For example, AAAUCUUCGU, if read in groups of three from the beginning, would be AAA, UCU, UCG, which are codons for lysine, serine, and serine, respectively. If the same sequence was read in groups of three starting from the second position, the groups would be AAU (asparagine), CUU (proline), and so on. The resulting amino acids would be completely different. For this reason, there are start and stop codons that indicate the beginning and ending of a sequence (or frame). AUG (methionine) is the start codon. UAA, UGA, and UAG, also known as ocher, opal, and amber, respectively, are stop codons.

Ribosomes synthesize proteins from mRNA in a process called translation. Sequences of three nucleotides called codons make up the strand of mRNA. Each codon codes for a specific amino acid. The ribosome is composed of two subunits, a larger subunit and a smaller subunit, which are composed of ribosomal RNA (rRNA). The smaller subunit of RNA attaches to the mRNA near the cap. The smaller subunit slides along the mRNA until it reaches the first codon. Then, the larger subunit clamps onto the smaller subunit of the ribosome. Transfer RNA (tRNA) has codons complementary to the mRNA codons. The tRNA molecules attach at the site of translation. Amino acids are joined together by peptide bonds. The ribosome moves along the mRNA strand repeating this process until the protein is complete. Proteins are polymers of amino acids joined by peptide bonds.

Decoding DNA Instructions

Two steps of protein synthesis decode DNA instructions: **Transcription** and **Translation**.

Transcription

Prokaryotic transcription consists of the following three stages:

1. Initiation: RNA polymerase binds to the start site on the RNA template strand, which is upstream in the promotor region of the gene.

2. Elongation: Synthesis of the new complementary RNA strand begins by the action of RNA polymerase as it works its way down the gene, growing the new strand.

3. Termination: The new RNA chain is completed and the RNA polymerase is released at the stop point, which is downstream on the gene in the terminator region.

Eukaryotic transcription is more complicated:

1. Eukaryotic Initiation: Eukaryotes contain advanced promoters, which are DNA sequences upstream of genes that recruit (or "bring in") transcription factors—proteins that facilitate or block RNA polymerase binding. The TATA box is where RNA polymerase II binds.

2. Eukaryotic Elongation: This is similar to prokaryotic elongation in that it generates a 5' to 3' complementary RNA transcript.

3. Eukaryotic Termination: A polyadenylation signal (DNA sequence AAUAAA) causes RNA polymerase to release several bases downstream. The RNA transcript is then processed. Upstream of the promoter, past a 5' untranslated region, a "cap" is added (like a telomere, but only a few repeated guanines). Downstream of the polyadenylation signal, a poly(A) tail is added. In addition to changes at the end of each transcript, areas within the transcript are further processed through splicing. Spliceosomes (large complexes of RNA and proteins) remove intermittent noncoding regions called introns, and the exons (coding sequences) are joined together. The introns are spliced out so that only the exons are part of the final transcript. Alternative splicing (when one transcript is spliced in different ways and creates different proteins) makes this process even more complicated.

Translation

Translation is the process of generating protein from RNA. Translation in prokaryotes is far simpler than in eukaryotes. They only have one circular chromosome (so far fewer genes), and they don't have DNA within a nucleus. They also don't process transcripts. Translation can even occur simultaneously with transcription on the same piece of RNA. In fact, many different ribosomes can be working on the same transcript at the same time, thus creating a structure called a **polyribosome**.

The ribosomes of bacteria and eukaryotes are similar. A ribosome has two subunits, both made of **ribosomal RNA** (**rRNA**, which is made by RNA polymerase I in eukaryotes). In eukaryotes, the small subunit and the large subunit assemble in conjunction with the AUG start codon in the mRNA message.

Before translation can start, a **transfer RNA** molecule (**tRNA**, which is made by RNA polymerase III) must join the ribosomal complex. tRNAs contain anticodons that are complementary to specific codons on the RNA transcript. On one side, their highly specific anticodon binds to corresponding codons in the mRNA. On the other side, tRNA molecules carry specific amino acids, the monomers of proteins. Once a tRNA molecule has released its amino acid, an enzyme aminoacyl-tRNA synthetase will join a free-floating tRNA and its corresponding amino acid. As a result, the tRNA can continue to deliver fresh amino acids to the ribosomes.

The point of translation is that, every time a tRNA molecule drops off an amino acid, it contributes to an emerging protein. Only when a stop codon is reached will the ribosome disassemble, thus releasing the assembled protein.

Second Base in Codon

First Base in Codon	U	C	A	G	Third Base in Codon
U	UUU } Phe UUC UUA } Leu UUG	UCU UCC } Ser UCA UCG	UAU } Tyr UAC UAA Stop UAG Stop	UGU } Cys UGC UGA Stop UGG Trp	U C A G
C	CUU CUC } Leu CUA CUG	CCU CCC } Pro CCA CCG	CAU } His CAC CAA } Gln CAG	CGU CGC } Arg CGA CGG	U C A G
A	AUU AUC } Ile AUA AUG Met or Start	ACU ACC } Thr ACA ACG	AAU } Asn AAC AAA } Lys AAG	AGU } Ser AGC AGA } Arg AGG	U C A G
G	GUU GUC } Val GUA GUG	GCU GCC } Ala GCA GCG	GAU } Asp GAC GAA } Glu GAG	GGU GGC } Gly GGA GGG	U C A G

Note in the chart above that each codon codes for a specific amino acid, even specifically coding for stop codons. There are many codon combinations, but only 20 amino acids. The redundancy in codon/amino acid pairs is due to the third base—or "**wobble**" position—of tRNA. The first two bases bind so strongly that sometimes the third base does not play much of an active role in connecting the codon and anticodon.

The ribosome has three tRNA landing spots. The first tRNA binds to the start codon in the "P" site, but every tRNA that follows lands at the "A" site. The growing amino acid chain is added to the amino acid in the A site (connecting via a peptide bond). The "E" site is where the naked tRNA exits after it has removed its amino acid chain.

Once a stop codon is reached, the amino acid chain leaves through an exit tunnel. At this point, it is an immature protein that is not properly folded and might be sent to the Golgi for modification.

Regulation of Expression

Transcriptional regulation in prokaryotes occurs via operons. An **operon** is a DNA sequence comprised of related genes which are clustered behind a single promoter that regulates them. Within the promoter is an operator, which is like an on/off switch.

Transcription Regulation

Eukaryotic transcription initiation involves transcription factors. Different transcription factors are present in different cell types, and certain sequences of DNA enhance protein binding. This results in an altered DNA spatial arrangement and exposure to RNA polymerase. Some transcription factors facilitate transcription while some block it. The presence or absence of these proteins in certain cells is one-way gene production is regulated.

Another way transcription is regulated in eukaryotes is by gene accessibility. DNA is wrapped around histone proteins in complexes called **nucleosomes**. These nucleosomes coil and supercoil, which make portions of the genome inaccessible. Modification of histone tails in nucleosomes open and close regions of DNA, making them either more or less available for the protein binding of transcription factors and RNA polymerase. When the chromatin is in its closed, coiled conformation, called *heterochromatin,* transcription is repressed because the genes are inaccessible. In contrast, in the *euchromatin* formation, the chromatin is open and uncoiled, which allows RNA polymerase to access the genes, which activates transcription.

Essentially, open chromatin (**euchromatin**) has acetylated histone tails with few methyl groups. **Heterochromatin** is the opposite, with histone tails being highly methylated and deacetylated so that DNA is closed and condensed.

MicroRNAs (**miRNAs**) and small interfering RNAs (**siRNAs**) can also regulate gene production. This is done by degrading certain transcripts or blocking their translation, and sometimes even altering chromatin structure.

It should be noted that there is far more DNA functionality than gene expression. DNA also contains promoters and enhancers to regulate gene expression, centromeres, telomeres, transposable elements, and other sequences.

Transposable Elements

These are literally "jumping genes" that relocate within an organism's genome. Prokaryotes and eukaryotes contain **transposons**. This supports the idea that they have a significant effect on biodiversity and arose through common ancestry among all living organisms. DNA sequences prone to transposon trespassing are predictable, repeated bases that can extend from 300 to 30,000 nucleotides.

Within an organism, transposons relocate and might leave the targeted segment behind. A transposon can make a copy of the DNA or simply cut out the sequence. Either way, transposase is a critical enzyme involved in the process.

Retrotransposons are elements that copy the targeted segment into an RNA transcript. The enzyme retrotransposase is then used to copy the intermediate back into DNA before inserting it, thus copying part of the genome and moving it elsewhere.

It's believed that all human ancestors once had brown eyes until a single individual had a mutation which coded for blue eyes. Once that allele entered the gene pool, it was passed to generations of offspring. Today it's fairly common to see a blue-eyed individual. Thus, changes in alleles can change phenotypes.

Variations are Introduced by the Imperfect Nature of DNA Replication and Repair

Despite replication's proofreading, mutations do happen. Mutations can be either point mutations or chromosomal mutations, and both increase genetic variation.

Point mutations involve a change in DNA sequence and are due to an addition, deletion, inversion, or substitution error.

Insertion and deletion mutations cause frame-shift mutations. This means that every following codon will be read incorrectly, massively changing the primary structure of the protein. Due to the redundancy of codon/amino acid pairing resulting from the wobble position, some substitution mutations cause no change in the protein sequence. The result is something called a silent mutation. All point mutations (including substitution and inversion) can cause gene malfunction.

Chromosomal Mutations

Sometimes there are **chromosomal mutations** in DNA replication and mitosis. The following types of mutations can occur:

- *Deletion*: A section of a chromosome is removed.
- *Duplication*: A section of a chromosome is repeated.
- *Inversion*: A chromosome is rearranged within itself.
- *Translocation*: Chromosome pieces mix or combine with other chromosomes.

Not only can mutations lead to changes in an individual's phenotype, but they can also contribute to reproductive isolation and speciation which have effects on a much larger scale.

In metaphase I in meiosis, homologous chromosomes align in the center of the cell. The law of independent assortment states that it is random whether a mother's or father's chromosomes go to one daughter cell or another. Since humans have 23 chromosomes, there are 2^{23} (over 8 million) possible variations. Independent assortment, random fertilization, and recombination contribute greatly to genetic diversity.

Molecular Diversity

Diversity exists in biology, even among items that are the same or similar, such as molecules, cells, and species. This diversity allows for a wider range of function and a greater ability to respond to the environment. One example of this is the production of antibodies by plasma cells. Antibodies are a class of proteins that provides a defense mechanism against pathogens. Their general structure is the same but each antibody has a slight variation at its tip that differentiates it for its specific pathogen. The plasma cells that make antibodies can make small changes in the antibody's DNA sequence to allow for the production of a large, diverse population of antibodies.

Organisms also have molecular diversity through genetic variation. One mechanism for creating genetic variation is by having two or more alternative forms, or alleles, of the same gene, such as with eye color or hair color. If both alleles are identical, the individual is considered homozygous; if the two alleles have different sequences, the individual is considered heterozygous. In most genes, one allele is considered more dominant than the other and will mask the appearance of the less dominant, or recessive, allele when there is a heterozygous situation. Another type of genetic variation that occurs is through the introduction of a mutation, which is a random, permanent alteration of a gene sequence. Some mutations that occur can be advantageous to a population and may help a population adapt to its environment. This may allow the organism to respond to stresses with more resilience and may even give them a fitness advantage. One example of a fitness advantage is in people whose hemoglobin gene is heterozygous for the sickle-cell trait. People who are homozygous for the sickle-cell trait have sickle-cell anemia, a harmful genetic disorder. People who are heterozygous for the sickle-cell trait are more resistant to malaria than people who are homozygous for wild-type, or normal, hemoglobin.

Gene Duplication

Genetic variation can also be established if a gene duplicates. Now the organism has an extra set of this gene, which gives the gene more chances to mutate since the original copies still maintain the necessary function of that gene. One example of this is the development of the antifreeze genes in fish that live in arctic environments, which prevents them from freezing. Originally, these fish had one copy of the sialic-acid synthase (*SAS*) gene. That gene was duplicated and, since the original *SAS* gene still retained its necessary function, the duplicated gene was no longer under selective pressure and could mutate. It then developed the mutations necessary to prevent fatal ice crystals from growing inside the fish.

Environmental Factors Influence Traits

Biological variation can also exist from environmental factors influencing gene expression. Two organisms with identical genomes can have different genes turned on or off depending on their different environments. The effect of the presence of lactose on bacteria that are *lac$^+$* is one example. *Lac$^+$* bacteria have a set of genes that metabolize the sugar lactose. When these bacteria are not around lactose, a repressor protein stops transcription of these genes. However, the presence of lactose causes the degradation of the repressor and the genes that metabolize lactose are transcribed. These environmental influences can also be seen at the organism level. Some cats have a temperature-sensitive mutation in a gene that is important for making fur color. At normal body temperature, this mutation disables the gene's function, so the fur is a light color. The cat's extremities are cooler and at a lower temperature, so the gene functions, giving the cat dark fur on its extremities.

Populations Need to Respond to Environmental Changes to Survive

The accumulation of these potential genetic variations can result in a diverse population. Populations with a high level of genetic diversity are able to respond better to changes in the environment, while populations with little genetic diversity are at risk for extinction.

One example of this is the destruction of potato crops in the mid-nineteenth century in Ireland because of a potato blight, which is a fungal infection. This destruction caused widespread famine and the death of hundreds of thousands of people. Potatoes can be grown by vegetative propagation, a type of asexual reproduction. The result of this asexual reproduction technique was that most of the potatoes in Ireland were genetic clones. All of these potatoes were highly susceptible to potato blight. If different strains of potatoes had been planted in Ireland, they would have had different levels of resistance to potato blight, and the famine caused by the loss of the potato plants would not have been as destructive.

Genetic diversity allows individuals in a population to respond differently to the same stimuli. This is seen with how individuals respond during a disease outbreak. Some individuals succumb to the disease, some are made ill, and others are not infected at all. These different responses are caused by the genetic makeup of the individual. Some individuals have genes that will prevent infection completely, others have immunological genes that can fight off infections, and still others have immunological genes that are unable to fight off the infection at all.

Allelic Variation Can be Modeled by the Hardy-Weinberg Equation

The different alleles that are present in a population can be modeled by the Hardy-Weinberg equilibrium. The Hardy-Weinberg equilibrium states that, absent any evolutionary influence, the proportion of alleles will remain constant throughout different generations. In cases where there are two different alleles, this equilibrium can be explained with the equation $p^2 + 2pq + q^2 = 1$, where p is the proportion of one allele in the population and q is the proportion of the other allele. In this equation, p^2 represents the number of homozygotes of one allele, $2pq$ represents the number of heterozygotes, and q^2 represents the number of homozygotes for the other allele.

Ecosystem Diversity

Ecosystem diversity is represented by the number of different species in the ecosystem. Ecosystems with higher diversity are more resilient to changes in the environment than simple ecosystems because the loss of any one species is not as detrimental.

The key factors in maintaining diversity in an ecosystem are keystone species, producers, and essential biotic and abiotic factors. Keystone species are any species whose role in the ecosystem is disproportionate to the size of the population. Although the keystone species may not be the most numerous or the most productive part of an ecosystem, its loss would devastate the ecosystem. For example, a keystone species may be a small predator that preys on an herbivorous species and keeps that species from eliminating all of a particular plant species. If the keystone species became extinct, the herbivorous species would completely wipe out the plant species and the ecosystem would change drastically. Similarly, in large bodies of water, the sea star is a keystone species that preys on sea urchins, which helps protect the coral reefs. Within an ecosystem, each species plays a specific, important role in preserving the environment they populate together.

Practice Questions

1. A color-blind male and a carrier female have three children. What is the probability that they are all color blind?
   a. 1/2
   b. 1/4
   c. 1/8
   d. 1/16
   e. 1/32

Use the image below to answer questions 2-3:

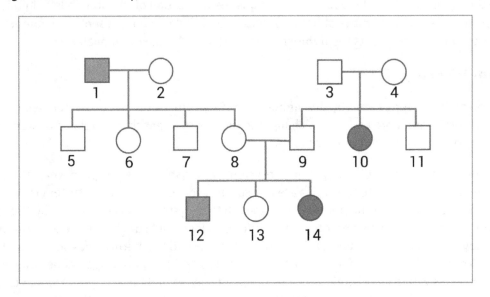

2. What kind of pedigree is shown?
   a. Autosomal dominant
   b. Autosomal recessive
   c. Sex-linked dominant
   d. Sex-linked recessive
   e. Heterozygous dominant

3. What is the genotype of individual 9?
   a. *AA*
   b. *Aa*
   c. *aa*
   d. $X^A Y$
   e. $X^a Y$

4. This karyotype indicates what about the individual?

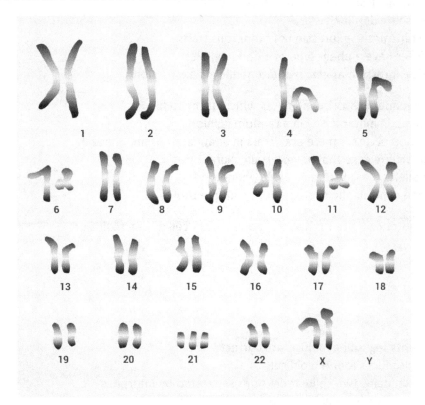

    I. They are female
    II. They are male
    III. They have Down Syndrome
    IV. They have Turner Syndrome

  a. Choices I and III
  b. Choices I and IV
  c. Choices II and III
  d. Choices II and IV
  e. Choices I, III, and IV

5. What is the probability of $AaBbCcDd \times AAbbccDD$ having a child with genotype $AAbbccDD$?
  a. 1/2
  b. 1/4
  c. 1/8
  d. 1/16
  e. 1/32

6. If a person with AB blood and a person with O blood have children, what is the probability that their children will have the same phenotype as either parent?
  a. 0%
  b. 25%
  c. 50%
  d. 75%
  e. 100%

7. Which of the following occurs in epistasis?
   a. There are two alleles: dominant and recessive
   b. Two alleles share dominance
   c. One gene controls the expression of numerous traits
   d. Heterozygotes have a phenotype of blended alleles
   e. One allele's expression affects a different allele's expression

8. The law of independent assortment states which of the following?
   a. Tetrads line up in metaphase I in a random fashion
   b. Nondisjunction occurs if there are errors in anaphase I or anaphase II
   c. Sperm will fertilize eggs that have certain characteristics
   d. Dominant alleles mask recessive allele phenotypes
   e. A single gene controls the expression of multiple traits

9. How many daughter cells are formed from one parent cell during meiosis?
   a. One
   b. Two
   c. Three
   d. Four
   e. Eight

10. Which statements regarding meiosis are correct?
    I. Meiosis produces four diploid cells.
    II. Meiosis contains two cellular divisions separated by interphase II.
    III. Crossing over occurs in the prophase of meiosis I.

    a. Choices I and II
    b. Choices I and III
    c. Choices II and III
    d. Choice III only
    e. Choice I only

11. Given the following recombination frequencies, which linkage map would be correct?
    D-A = 9%
    D-B = 3%
    B-C = 5%
    C-A = 1%
    a. ACBD or DBCA
    b. DCAB or BACD
    c. BADC or CDAB
    d. ABCD or DCBA
    e. BADC or DBCA

12. Which of the following is true regarding the *lac* operon in prokaryotes?
    a. It is repressible
    b. It is blocked by a repressor that binds to the *lacZ* site in the absence of lactose
    c. It regulates one gene important for lactose digestion
    d. It is also regulated by tryptophan
    e. It is also regulated by the presence of glucose and cAMP

13. Which of the following sequences are significant during transcription of eukaryotic mRNA?
    I. TATA
    II. AAAUAA
    III. AUG

    a. Choice I only
    b. Choice II only
    c. Choice III only
    d. Choices I and II
    e. Choices I, II, and III

14. In protein synthesis, which of the following is NOT true about the molecule tRNA?
    a. It obtains its amino acid by aminoacyl-tRNA synthetase
    b. It has an anticodon that is complementary to an mRNA codon
    c. The wobble position is the third base in the anticodon
    d. It contains a large subunit and a small subunit
    e. It is synthesized by RNA polymerase III

15. DNA must be created in a 5' to 3' direction. This causes which of the following?
    a. Shortening of telomeres
    b. Hydrogen bonds between nitrogen bases
    c. Primase binding
    d. Okazaki fragments to form on the leading strand
    e. RNA polymerase bonding

16. Which phenomenon is *not* involved in prokaryotic genetic diversity?
    a. Mutation
    b. Transformation
    c. Conjugation
    d. Transduction
    e. Crossing over

17. Which statement is NOT true regarding the lytic and lysogenic cycles?
    a. Both processes begin with phage attachment.
    b. Only the lysogenic cycle applies to eukaryotic cells.
    c. Lysogenic is dormant until an activating factor stimulates viral production.
    d. Both cycles utilize host cell ribosomes to make more viruses.
    e. The lysogenic cycle involves a prophage—phage DNA integrated with host cell DNA.

18. Which event occurs first in receptor-mediated signal transduction involving receptors?
    a. Phosphorylation/dephosphorylation cascade
    b. Transcription factor activation by activated protein
    c. Ligand binding to a tyrosine kinase transmembrane protein
    d. Ion-gated channels open
    e. Adenylyl cyclase converts ATP to cAMP

19. With which genotype would the recessive phenotype appear, if the dominant allele is marked with "A" and the recessive allele is marked with "a"?

    a. AA

    b. aa

    c. Aa

    d. aA

    e. It cannot be determined with the information given.

20. Blood type is a trait determined by multiple alleles, and two of them are co-dominant: I^A codes for A blood and I^B codes for B blood. The i allele codes for O blood and is recessive to both. If an A heterozygous individual and an O individual have a child, what is the probably that the child will have A blood?

    a. 0%

    b. 25%

    c. 50%

    d. 75%

    e. 100%

Answer Explanations

1. B: 1/8. Color blindness is a recessive, sex-linked trait. The Punnett square below shows the cross between a carrier female and a color-blind male. The two offspring in bold are color blind. The probability of having one child that is color blind is 1/2. The probability of having three color-blind children is $\frac{1}{2} \times \frac{1}{2} \times \frac{1}{2}$ (law of multiplication) or 1/8.

	X^C	X^c
X^c	$X^C X^c$	$\mathbf{X^c X^c}$
Y	$X^C Y$	$\mathbf{X^c Y}$

2. B: Autosomal recessive. For dominant pedigrees, it would be impossible for two recessive parents to have a child that expresses the dominant trait, as is seen in individual 10, making Choices *A* and *C* wrong. This cannot be a sex-linked recessive pedigree because of individual 14: a girl cannot be color blind unless her father is color blind, making Choice *D* incorrect. The correct answer is *B*, autosomal recessive.

3. B: *Aa*. This is an autosomal recessive pedigree, so Choice *D* and *E* are incorrect. Individual 9 has a child who has the trait, so he must have a recessive allele. He must also have the dominant allele since he does not have the trait. Choice *B* is the heterozygous genotype that has both the dominant and recessive allele, so the correct answer is *B*.

4. A: I and III. This is a female because her 23rd chromosome pair is composed of two X chromosomes and no Y. The karyotype also shows trisomy 21, which is Down syndrome. Turner syndrome is monosomy 23 (women with only one sex chromosome), making IV incorrect.

5. D: 1/16. The probability of each specified genotype can be determined by individual Punnett squares. Each probability should then be then multiplied (law of multiplication) to find the value, which in this case is $\frac{1}{2} \times \frac{1}{2} \times \frac{1}{2} \times \frac{1}{2} = \frac{1}{16}$.

	A	a
A	AA	Aa
A	AA	Aa

Probabilities: *AA* = 1/2

	B	b
b	Bb	bb
b	Bb	bb

bb = 1/2

	C	C
c	Cc	cc
c	Cc	cc

cc = 1/2

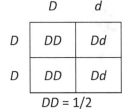

	D	d
D	DD	Dd
D	DD	Dd

DD = 1/2

6. A: 0%. There is no chance that an offspring will be O blood or AB blood (see Punnett square).

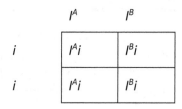

7. E: One allele's expression affects a different allele's expression. Co-dominance and incomplete dominance are described in Choices *B* and *D*, not epistasis, so they are incorrect. Simple Mendelian genetics is described in Choice *A*, which is also incorrect. Choice *C* describes pleiotropy, so it is incorrect.

8. A: Tetrads line up in metaphase I in a random fashion. Choice *B* is not a heredity issue; nondisjunction is a mistake in chromatid separation that is not inherited. Choice *C* is untrue because fertilization is random. Choice *D* explains alleles, but does not explain the mechanism behind genetic diversity like Choice *A* does. The law of independent assortment pertains to the random lineup of chromosomes in metaphase I. Choice *E* is incorrect as it describes pleiotropy.

9. D: Meiosis has the same phases as mitosis, except that they occur twice—once in meiosis I and once in meiosis II. During meiosis I, the cell splits into two. Each cell contains two sets of chromosomes. Next, during meiosis II, the two intermediate daughter cells divide again, producing four total haploid cells that each contain one set of chromosomes.

10. D: Choice III only. Choice I is incorrect because meiosis produces haploid cells. Choice II is incorrect because there is no interphase II (otherwise gametes would be diploid instead of haploid). Choice *D* is the only correct answer because the others contain choices I and II.

11. A: ACBD or DBCA. This order is the only one that works so the map units have the assigned distances between them.

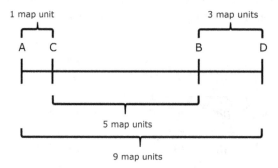

D-A = 9%

D-B = 3%

B-C = 5%

C-A = 1%

12. E: It is also regulated by the presence of glucose and cAMP. Operons in bacteria have a regulatory "switch" (DNA sequence called an operator) that controls transcription of several related genes, so Choice *C* is incorrect. There are three genes for lactose digestion, not one. The *lac* operon is inducible. It is mostly off and only turned on by the presence of lactose, so *A* is incorrect. Similarly, because it is inducible, lactose will bind to the repressor so that RNA polymerase is no longer blocked, so Choice *B* is incorrect. Choice *D* is incorrect because tryptophan is unrelated to the *lac* operon. The only correct description of the *lac* operon is Choice *E*.

13. D: Choices I and II. Choice I, the TATA box, is important for RNA polymerase recruitment. Choice II (AAAUAA) is the poly(A) tail that is a signal to stop transcription. Choice III (AUG) is the start codon, which is an important sequence for translation, not transcription. Therefore, only the first two sequences are important during mRNA transcription.

14. D: It contains a large subunit and a small subunit. rRNA is the type of RNA that has subunits, not tRNA. The other answer choices (*A*, *B*, *C*, and *E*) are all true.

15. A: Shortening of telomeres. Choices *B*, *C*, and *E* all occur, but they have nothing to do with the direction of DNA synthesis. Choice *D* is untrue because Okazaki fragments are formed due to the directional synthesis of DNA, but they form on the lagging strand.

16. D: Crossing over. All choices cause genetic variation in bacteria except for crossing over, which is strictly a eukaryotic event that occurs with linear, homologous chromosomes.

17. B: Only the lysogenic cycle applies to eukaryotic cells. Choice *C* is true since the lysogenic cycle is dormant until an activating factor stimulates viral production. Choice *A* is true because both start with viral attachment. Choice *D* is true because the lysogenic cycle, once activated, utilizes host cell membranes like the lytic does. Choice E is true because the lysogenic cycle does involve a prophage. Choice *B* is untrue because the lysogenic cycle is specific to prokaryotes and is not a process in eukaryotic viral infections, thus Choice *B* is the correct answer.

18. C: Ligand binding to a tyrosine kinase transmembrane protein. Tyrosine kinase transmembrane proteins are just one example of a receptor protein. There are also G protein-coupled receptors and ion channel receptors. Regardless of the type of receptor, ligand binding is the first step. Choice *D* is incorrect because, prior to ion-gated channels opening, a ligand would need to bind to the receptor to induce the conformational change. Choice *A* occurs after ligand binding and Choice *B* is a response that is much farther downstream (it's an effect of signal transduction). Choice *E* is subsequent to the activation of a G protein-coupled receptor by signaling molecules such as epinephrine.

19. B: Dominant alleles are considered to have stronger phenotypes and, when mixed with recessive alleles, will mask the recessive trait. The recessive trait would only appear as the phenotype when the allele combination is *aa* because a dominant allele is not present to mask it.

20. B: 50%. According to the Punnett square, the child has a 2 out of 4 chance of having A-type blood, since the dominant allele I^A is present in two of the four possible offspring. The O-type blood allele is masked by the A-type blood allele since it is recessive.

$I^A i$	ii
$I^A i$	ii

Organismal Biology

Structure, Function, and Development of Organisms

Basic Anatomy and Physiology of Animals, Including the Human Body

Response to Stimuli and Homeostasis

A **stimulus** is a change in the environment, either internal or external, around an organism that is received by a sensory receptor and causes the organism to react. **Homeostasis** is the stable state of an organism. When an organism reacts to stimuli, it works to counteract the change in order to reach homeostasis again.

Exchange with the Environment

Animals exchange gases and nutrients with the environment through several different organ systems. The **respiratory system** mediates the exchange of gas between the air and the circulating blood, mainly through the act of breathing. It filters, warms, and humidifies the air that gets inhaled and then passes it into the blood stream. The main function of the **excretory system** is to eliminate excess material and fluids in the body. The kidneys and bladder work together to filter organic waste products, excess water, and electrolytes from the blood that are generated by the other physiologic systems, and excrete them from the body. The **digestive system** is a group of organs that work together to transform ingested food and liquid into energy, which can then be used by the body as fuel. Once all of the nutrients are absorbed, the waste products are excreted from the body.

Internal Transport and Exchange

The **circulatory system** is composed of the heart and blood vessels. The **heart** acts as a pump and works to circulate blood throughout the body. Blood circulates throughout the body in a system of vessels that includes arteries, veins, and capillaries. It distributes oxygen, nutrients, and hormones to all of the cells in the body. **Arteries** transport oxygen-rich blood from the heart to the rest of the tissues in the body. The largest artery is the **aorta**. **Veins** collect oxygen-depleted blood from tissues and organs and return it back to the heart. **Capillaries** are the smallest of the blood vessels and do not function individually. Instead, they work together in a unit—called a **capillary bed**—to transport both oxygen-rich and oxygen-poor blood to other vessels.

Control Systems

The **nervous system** is one of the smallest but most complex organ systems in the human body. It consists of all of the neural tissue and is in charge of controlling and adjusting the activities of all of the other systems of the body. It is divided into the **central nervous system (CNS)** and the **peripheral nervous system (PNS)**. The CNS is where intelligence, memory, learning, and emotions are processed. It is responsible for processing and coordinating sensory data and motor commands. The PNS is responsible for relaying sensory information and motor commands between the CNS and peripheral tissues and systems.

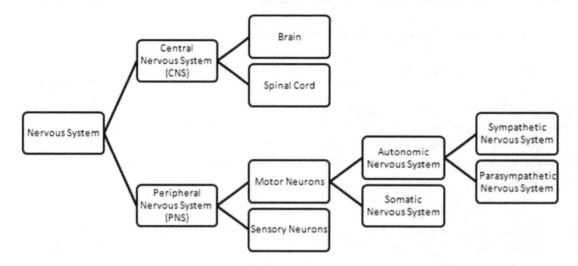

The **endocrine system** is made up of the ductless tissues and glands that secrete hormones into the **interstitial fluids** of the body, which are the fluids that surround tissue cells within the body. This system works closely with the nervous system to regulate the other physiologic systems in order to maintain homeostasis. It acts by releasing hormones into the bloodstream, which are then distributed to the whole body.

Movement and Support

The adult **skeletal system** consists of the 206 bones that make up the skeleton, as well as the cartilage, ligaments, and other connective tissues that stabilize the bones. It provides structural support for the entire body, a framework for the soft tissues and organs to attach to, and acts as a protective barrier for some organs, such as the ribs protecting the heart and lungs, and the vertebrae protecting the spinal cord.

The **muscular system** is responsible for all body movement that occurs. Body movements occur by muscle contractions that cause specific actions or joint movements. There are approximately seven hundred muscles in the body that are attached to the bones of the skeletal system. As mentioned, there are three types of muscle tissue: skeletal, cardiac, and smooth. **Skeletal muscles** are voluntary muscles that attach to bones through tendons. **Cardiac muscle tissue** is found only in the heart and is involuntary. **Smooth muscle tissue** lines the walls of hollow structures, such as blood vessels, the stomach, and the bladder, and is involuntary. When smooth muscle tissue contracts, the structure it lines narrows or constricts.

Reproduction and Development

The **reproductive system** is responsible for producing and maintaining functional reproductive cells in the human body. The human male and female reproductive systems are very different from each other. The male gonads, called **testes**, mainly secrete testosterone, which is a steroid hormone that controls the development and maintenance of male physical characteristics. They also produce **sperm cells**, which are responsible for fertilizing the female reproductive cell in order to produce offspring. The female gonads, called **ovaries**, generally produce one immature egg per month. They are also

responsible for secreting the hormones estrogen and progesterone. Once an egg is fertilized by a sperm cell, an **embryo** develops. After approximately 40 weeks of gestation, a baby is born.

Accomplishment of Regulation with Feedback Loops

Feedback loops are complex regulators that keep a balance between energy expenditure and energy conservation. One example of a feedback loop includes the complex mechanisms that maintain a human's optimal body temperature. This intricate feedback loop is controlled by the hypothalamus—the link between the nervous system and endocrine system in the brain. Upon receiving a neural stimulus, the hypothalamus stimulates or inhibits the pituitary gland based on nervous system input, which is how it regulates hormone levels that have specific physiological effects. When body temperature is too low, the body is prompted to shiver, which releases energy from the muscles and warms the body. When body temperature is too high, the body initiates sweating, which regulates temperature through a process known as evaporative cooling. Evaporative cooling occurs because the water droplets heated by the body and released by sweat glands are the ones with the highest kinetic energy. The high kinetic energy results in the evaporation of the hottest water molecules, which leaves the cooler ones behind. Other endotherms, or organisms that regulate their temperature internally, have different mechanisms to cool down. Dogs, for example, only have sweat glands in their paws and nose, so they pant to use evaporative cooling by drawing moisture from their lungs.

Endothermy Versus Ectothermy for Temperature Regulation

Endotherms, such as humans, regulate body temperature internally. Conversely, ectotherms, such as reptiles, have no internal mechanism for temperature regulation. They bask in the sun or hide in the shade to regulate their body temperature. They have a significantly slower metabolism and require far less food than endotherms because they do not require advanced mechanisms to regulate temperature.

Reproduction Requires Energy

Sustaining life through energy conservation would be futile if there were no life to begin with. Every organism on the planet exists because of reproduction. In every known species on Earth, there is an inherent drive to reproduce, which requires a great deal of energy. If organisms do not expend energy procreating and ensuring the survival of their offspring, they are in danger of not passing on their genes. However, an organism cannot reproduce if it does not survive itself, so it must prioritize survival over reproduction. If survival seems unlikely, an animal will expend energy to fight or flee before it will reproduce.

For example, many bird species have elaborate courtship rituals that take a significant amount of energy in the form of dancing, squawking, and jumping around. If a male bird is about to seduce a female, but a predator approaches, he will use any available energy to flee. Energy at that time is best used to preserve his life. However, if no predator is present, he will use this energy to attempt to reproduce.

Hibernation as a Reproduction and Energy Conservation Strategy

Hibernating animals are another example, since they are generally out of reproductive commission in the winter. Their internal temperature drops to around freezing, and their metabolic rate decreases as much as 20 times lower than during arousal. This is to their benefit. They spend their limited energy stores solely on personal survival and abandon any actions toward procreation and nurturing. Similar to hibernation, estivation is the opposite trend with the same effect. This is when animals retreat in the summer to avoid the heat and water scarcity, and in doing so, they disregard any reproductive responsibilities.

The Effect of Body Size on Metabolic Rate

The metabolic rate is the amount of energy used by an organism to perform reactions in a given time period. One might assume that the larger the organism, the higher the metabolic rate, but actually, the size of the organism is inversely proportional to its metabolic rate. Small organisms have a greater surface area to volume (SA:V) ratio than larger organisms. The surface area is the total area of the outermost layer of an object, and is the area through which organisms lose heat. To illustrate this point, think of a sphere. The volume of a sphere ($4 / 3\pi r^3$) is much greater than its surface area ($4\pi r^2$), and increases at a much faster relative rate. With a huge sphere, there's a lot more matter inside the sphere than on the outermost layer of it. However, with a small sphere, there isn't as much matter on the inside compared to its surface layer. Therefore, the larger organisms become, the smaller the SA:V ratio. Smaller organisms have a larger SA:V ratio and hence, lose heat much faster than larger organisms. The greater metabolic rate of smaller organisms is required in order to maintain a constant body temperature.

Immune System

The **immune system** comprises cells, tissues, and organs that work together to protect the body from harmful foreign invaders. It is important for the body's immune system to be able to distinguish between **pathogens**, such as viruses and parasites, and the body's own healthy tissue. There are two types of immune systems that work to defend the body against infection: the innate system and the adaptive system. The **innate immune system** works without having a memory of the pathogens it defended against previously. The **adaptive immune system** creates a memory of the pathogen that it fought against, so that the body can respond again in an efficient manner the next time the pathogen is encountered. When **antigens** or **allergens** such as pollen are encountered, antibodies are secreted to inactivate the antigen and protect the body.

If the immune system is not functioning properly, the body may develop an **autoimmune disorder**. In this case, the body cannot distinguish between itself and foreign pathogens, so it attacks itself unnecessarily.

Changes in Structure Can Cause Changes in Function

Interactions between molecules are the driving force behind biological processes. The structures of the different molecules facilitate their interaction. Biological processes progress because this interaction causes a change in the structure of one or more of the molecules, facilitating interactions with new molecules.

The Effects of Enzyme Shapes

Enzymes are proteins that catalyze a reaction, which means they make a reaction happen faster. Almost all biological reactions are catalyzed by enzymes. In order for the enzyme to function, it needs to interact with the substrates, which are the components of the reaction. This process is very precise to ensure that only the correct substrates are involved in the reaction. Therefore, the substrates have a very specific fit into the active site of the enzyme, which is the area that catalyzes the reaction. Once the enzyme and substrate bind to the active site, they form what is called the **enzyme-substrate complex**.

Cofactors and Coenzymes

To function correctly, some enzymes require cofactors, such as vitamins, and coenzymes. These molecules affect the shape of the enzyme. When they are not present, the substrate cannot fit into and

interact with the enzyme. However, when the cofactors or coenzymes are present, the shape of the enzyme changes, and it can interact with the substrate.

Because this fit needs to be so perfect, it offers a method for regulation of enzyme activity. Other factors or molecules may bind to the enzyme and change its shape. This alters the enzyme's ability to bind to the substrates. These molecules bind to allosteric sites, which are sites that are different from the active site. This binding can either change the shape to allow the substrate to fit, which is called allosteric activation, or it can prohibit the substrate from fitting, which is called allosteric inhibition.

Major Structures of Plants and Their Functions

Characteristics of Vascular and Nonvascular Plants
Plants that have an extensive vascular transport system are called **vascular plants**. Those plants without a transport system are called **nonvascular plants**. Approximately ninety-three percent of plants that are currently living and reproducing are vascular plants. The cells that comprise the vascular tissue in vascular plants form tubes that transport water and nutrients through the entire plant. Nonvascular plants include mosses, liverworts, and hornworts. They do not retain any water; instead, they transport water using other specialized tissue. They have structures that look like leaves, but are actually just single sheets of cells without a cuticle or stomata.

Structure and Function of Roots, Leaves, and Stems
Roots are responsible for anchoring plants in the ground. They absorb water and nutrients and transport them up through the plant. **Leaves** are the main location of photosynthesis. They contain **stomata**, which are pores used for gas exchange, on their underside to take in carbon dioxide and release oxygen. **Stems** transport materials through the plant and support the plant's body. They contain **xylem**, which conducts water and dissolved nutrients upward through the plant, and **phloem**, which conducts sugars and metabolic products downward through the leaves.

Asexual and Sexual Reproduction
Plants can generate future generations through both asexual and sexual reproduction. Asexually, plants can go through an artificial reproductive technique called **budding**, in which parts from two or more plants of the same species are joined together with the hope that they will begin to grow as a single plant.

Sexual reproduction of flowers can happen in a couple of ways. **Angiosperms** are flowering plants that have seeds. The flowers have male parts that make pollen and female parts that contain ovules. Wind, insects, and other animals carry the pollen from the male part to the female part in a process called **pollination**. Once the ovules are pollinated, or fertilized, they develop into seeds that then develop into new plants. In many angiosperms, the flowers develop into fruit, such as oranges, or even hard nuts, which protect the seeds inside of them.

Nonvascular plants reproduce by sexual reproduction involving **spores**. Parent plants send out spores that contain a set of chromosomes. The spores develop into sperm or eggs, and fertilization is similar to that in humans. Sperm travel to the egg through water in the environment. An embryo forms and then a new plant grows from the embryo. Generally, this happens in damp places.

Growth
Germination is the process of a plant growing from a seed or spore, such as when a seedling sprouts from a seed or a sporeling grows from a spore. Plants then grow by **elongation**. Plant cell walls are

modified by the hormone auxin, which allows for cell elongation. This process is regulated by light and phytohormones, which are plant hormones that regulate growth, so plants are often seen growing toward the sun.

Uptake and Transport of Nutrients and Water

Plant roots are responsible for bringing nutrients and water into the plant from the ground. The nutrients are not used as food for the plant, but rather to maintain the plant's health so that the plant can make its own food during photosynthesis. The xylem and phloem in the stem help with transport of water and other substances throughout the plant.

Responses to Stimuli

Because plants have limited mobility, they often respond to stimuli through changes in their growth behavior. **Tropism** is a response to stimuli that causes the plant to grow toward or away from the stimuli:

- **Phototropism**: A reaction to light that causes plants to grow toward the source of the light

- **Thermotropism**: A response to changes in temperature

- **Hydrotropism**: A response to a change in water concentration

- **Chemotropsim**: Growth or movement in response to chemicals

- **Gravitropism**: A response to gravity that causes roots to follow the pull of gravity and grow downward, but also causes plant shoots to act against gravity and grow upward

Animal Behavior

Behavior is the way in which an organism responds to a stimulus. It can range from turning toward a sound or birds migrating to specific locations at certain times of the year. The sound reaction is an obvious stimulus, but migration is a more complicated stimulus. Studies have shown that day/night circadian clocks, the position of the sun, and even the Earth's magnetic field are interacting stimuli responsible for migratory behavior.

Behavior is not only a response to the environment, but also involves communication and relationships between organisms. Birds communicate to each other in elaborate dances or songs. Humans communicate to each other through language. Fruit fly communication provides an example that demonstrates visual, chemical, tactile, and auditory animal communication, examples of which are outlined below.

- Visual communication: Male fly spies a female and orients his body so that she can see him.
- Chemical communication: Male fly smells a female.
- Tactile communication: Male fly approaches a female and pokes her with his leg.
- Auditory communication: Male fly produces a song via wing vibrations and seduces a female.

Pheromones are chemicals that produce an olfactory stimulus between organisms in a species. Queen bees secrete a pheromone that inhibits ovarian development in her workers and attracts males when she wants to mate, ensuring that her progeny are the only bees to survive.

Innate Behaviors are Inherited

Innate behavior is quite different than behavior based on learning. Learning results in behavior modification derived from experiences. There are some behaviors that are both innate and learned, such as imprinting, which involves a specific period of time during development where an organism is susceptible to learning. The classic example of imprinting focuses on baby geese. There is a period of time not only where the geese offspring mimics its parents and learns species behavior, but also where the parent learns to recognize its offspring. If geese do not bond with their offspring in the first few days, the parents won't care for their infants.

Learning Occurs Through Environmental Interactions

Spatial learning is demonstrated by birds that can return to their nests after hunting, cats that learn where their litter box is, and squirrels that remember where they buried acorns.

More advanced learning involves cognition and problem-solving, and these high-level processes are not just reserved for humans. Rats in mazes and chimpanzee sign language acquisition provide evidence for other mammalian cognition. Even insects have exhibited problem-solving in some studies.

Animal Behaviors are Triggered by Environmental Cues

Behavior is responding to environmental information/stimuli such as phototropism and photoperiodism in plants discussed previously. Animals respond to environmental cues as well. For example, the males in many species of birds have brighter feathers than the females, which is often achieved by the pigments of the fruit in their diet that are deposited into their feathers. The brighter the bird's feathers, the richer and more frequent his diet. Female birds spot these color differences and will choose the male with the brighter colors over the others — ensuring his reproductive success. Her behavior, on the other hand, is in response to his courtship rituals.

Cooperative behavior in ecosystems is also adaptive and contributes to survival. Mutualism, such as protozoa living in a termite's gut, benefits both organisms; the termite gets aid in digesting wood that it cannot get on its own, and the protozoa get protection with easily obtained nutrients in exchange.

Animals Respond to Environmental Cues

Animals also respond to environmental cues in a synchronous manner. Circadian rhythms are the predictable animal responses to 24-hour days and are partially controlled by melatonin, a hormone that aids in sleep. High melatonin expression occurs in winter, a period of long nights.

Women's menstrual cycles are also cyclical and regulated by tight hormone control. The cycle begins with the release of a hormone called gonadotropin-releasing hormone (GnRH) by the hypothalamus, which causes the anterior pituitary to secrete low levels of FSH (follicle-stimulating hormone) and LH (luteinizing hormone). FSH stimulates follicular growth in the follicular phase, and the follicles begin to produce estradiol. Oocytes, or eggs, mature, and estradiol feeds back to the hypothalamus/pituitary, inhibiting FSH and LH expression. Pre-ovulation, estradiol spikes to stimulate the hypothalamus/pituitary so that GnRH, FSH, and LH levels significantly increase. Ovulation occurs when the follicle releases the egg. At this point, the follicular phase ends and the luteal phase begins. The left-behind follicular tissue transforms into a corpus luteum, which responds to high LH levels by secreting progesterone. The progesterone inhibits hypothalamic/pituitary production of LH and FSH so that another egg does not mature. The corpus luteum disintegrates and reduces progesterone and estradiol production. These low steroid levels free the hypothalamus and pituitary to repeat the 28-day cycle.

While the egg is maturing, the hormones are preparing the uterus for pregnancy in case the egg is fertilized. Estradiol and progesterone control uterine development. Post-ovulation, estradiol and progesterone stimulate development of the uterine lining by increasing blood flow and glandular development for nutrient attainment. When the corpus luteum degrades and these hormones decline, the uterine wall disintegrates and sloughs off along with blood during menses. Once the follicle develops and estradiol production increases, the uterine lining is developed again, continuing the cycle.

Practice Questions

1. The hypothalamus stimulates the pituitary gland with thyrotropin-releasing hormone (TRH), and the pituitary stimulates the thyroid gland by producing thyrotropin—also known as thyroid-stimulating hormone (TSH). Upon activation of the thyroid, hormones T3 and T4, which maintain and stimulate metabolism, are released. TSH also stimulates the thyroid to release calcitonin, which lowers calcium in the blood. Which of the statements below is the most likely regulatory mechanism that fine-tunes metabolism?

    a. As blood calcium levels increase, TSH production also increases.

    b. T_3 and T_4 increase the secretion of calcitonin.

    c. The more TSH released, the more calcium in the blood.

    d. TSH and calcium levels are independent of each other because they have different targets.

    e. As levels of TRH increase, the body's temperature decreases.

2. A study that investigated respiration rates of different organisms recorded the data below. Use the data to answer the question below.

	Hours required for 1 g of the animal to use 10 mL O_2
Bird	1.3
Human	2.7
Cat	1.9
Elephant	8.3
Lizard	25.2

Which statement best describes the dramatic difference in the data between the lizard and other organisms?

    a. The small size of lizards means that it will have the lowest metabolic rate due to its surface area to volume ratio.

    b. Mitochondria structure is different in endotherms and ectotherms, which results in dramatically different metabolic rates.

    c. The small size of lizards means that it will have the greatest metabolic rate due to its surface area to volume ratio.

    d. Lizards require much less energy to calibrate their temperature because of their habitats.

    e. Internal homeostatic mechanisms for temperature regulation are vastly different between lizards and the other animals.

3. Two different plants were grown in a lab and their response to light was investigated. Based on the representative qualitative data shown in the diagram, which of these mechanisms would best explain the flowering patterns?

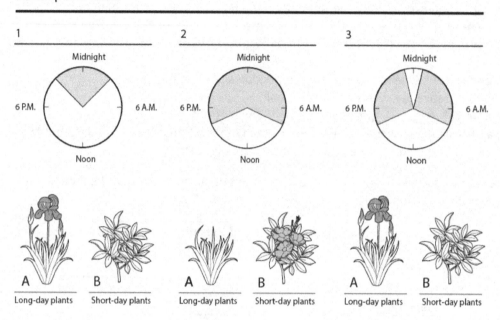

Photoperiodism

a. Auxin's property of elongating cells exposed to light is responsible for flowering in both plants.
b. Plant B has a selective advantage because it is unusually reproductively active in the winter, which reduces competition.
c. Plant A requires a threshold of sunlight in order to flower. Its blooming is independent of the season.
d. Both plants are dependent on the amount of continuous light exposure in order to flower.
e. Plant A has a selective advantage because it flowers as long as it receives a burst of light at midnight, making its requirements for survival easily satisfied.

4. The image below shows hormone levels during a 28-day human menstrual cycle. Days 1-14 are when the follicle develops, and day 14 is ovulation and when the egg is released. During days 15-28, the follicle left behind becomes the corpus luteum. This also is the time when the uterine lining vascularizes and develops.

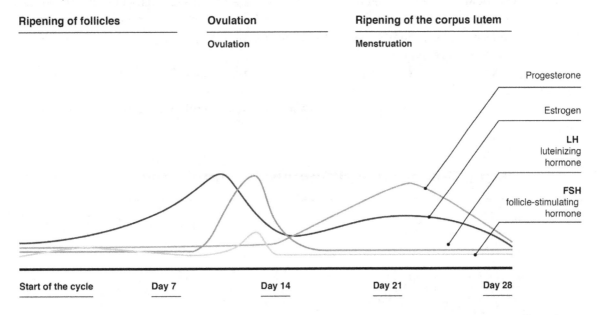

The interrelationship between the hormones is critical in the regulation of the menstrual cycle. Based on the image, which is the most likely and consistently demonstrated conclusion regarding hormone regulation?
   a. FSH is required for progesterone production.
   b. Estrogen is required for LH production.
   c. Progesterone is required for estrogen, LH, and FSH production
   d. FSH and LH require the same negative feedback loop.
   e. LH stimulates estrogen release.

5. One reason for women's elaborate control of egg release is that there is a finite number of eggs. Women's eggs are distributed very slowly throughout her fertility window due to the tight hormonal regulation noted in the previous question. How is a woman's energy expenditure benefited by her hormonal regulation?
   a. If the uterine lining were continuously vascularized and ready for egg implantation, metabolic cellular processes would be too costly from an energy perspective.
   b. Menses are preceded by a few days of low energy, where some women become very tired and decrease activity. This energy conservation helps prepare for the next egg release.
   c. The hormonal release simultaneously releases pheromones, which attract men during ovulation and is incidentally women's most fertile time of their cycle.
   d. It is beneficial to release the egg once a month to ensure only one egg is fertilized so that pregnancy does not drain the body of too many resources.
   e. Because energy expenditure decreases during menstruation, pregnant women are in a metabolically advantageous position to gain necessary weight for a healthy fetus.

6. The human immune system has an arsenal of white blood cells. The following image shows the sequence of events in the humoral immune response. How would this process be affected if the individual had been vaccinated against the pathogen prior to infection?

a. Antibodies would be freely circulating and cause a more efficient secondary immune response.

b. T cells would be circulating and begin to produce antibodies much quicker due to the vaccination.

c. Existing B cells will quickly differentiate into plasma cells upon cytokine stimulation, causing a more robust and faster response.

d. Phagocytic lymphocytes are already presenting MHC class II molecules that activate the immune response.

e. Immunoglobulins would be freely circulating and cause a more robust and faster secondary immune response.

Clonal Selection and Ensuing Events

1

Antigen recognition

Immunocompetent B cells exposed to antigen. Antigen binds only to B cells with complementary receptors.

2

Antigen presentation

B cell internalizes antigen and displays processed epitope. Helper T cell binds to B cell and secretes interleukin.

3

Clonal selection

The Interleukin stimulates the B cell to make it divide repeatedly and form a clone.

4

Differentiation

Some cells of the clone become memory B cells. Most differentiate into plasma cells.

5

Attack

The plasma cells synthesize and secrete antibodies. These antibodies then employ various means to render antigens harmless.

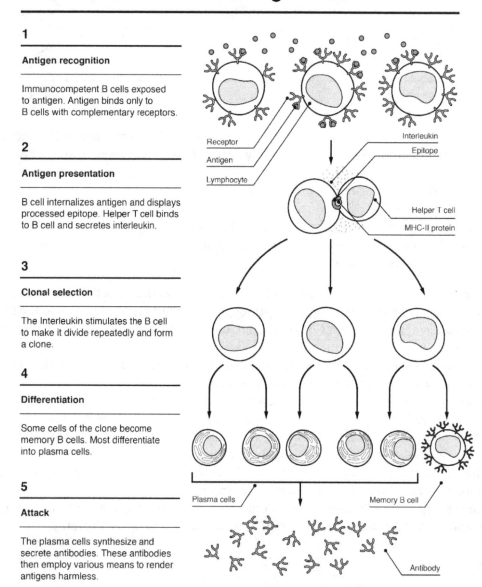

7. Microvilli in the small intestine serve an important function by increasing nutrient absorption. The three conditions in the table all affect blood sugar circulation in different ways, which are all indirectly related to absorption. The more efficient the absorption, the more sugar enters the bloodstream. Type I diabetes involves little to no insulin production, which causes very high blood sugar if not treated with exogenous insulin. People with celiac disease have trouble digesting gluten. Hypoglycemia results in very low blood sugar.

Microvilli Size in Different Groups of Individuals		
	Average Length (μm)	Average Width and Height (μm)
Unaffected	4.7	0.8
Type I diabetes	3.3	1.7
Celiac	1.1	1.7
Hypoglycemia	5.0	0.5

Which statement is a reasonable conclusion given the data recorded? The sample size was very small, and although these numbers do not necessarily reflect that of the whole population, assume that they do.

a. The individuals with celiac disease have the smallest surface area to volume ratio, which will be beneficial because it will make nutrient absorption more efficient.

b. The individuals with hypoglycemia have a larger surface area to volume ratio, which will be beneficial because it will make nutrient absorption more efficient.

c. The unaffected individuals have the largest surface area to volume ratio, which will be beneficial because it will make nutrient absorption more efficient.

d. The unaffected individuals have the smallest surface area to volume ratio, which will be beneficial because it will make nutrient absorption more efficient.

e. The individuals with celiac disease have the largest surface area to volume ratio, which will be beneficial because it will make nutrient absorption more efficient.

8. A characteristic of life is that organisms are able to react and respond to their environment. The choices below are proposed physiological regulatory responses of various organisms. Which one actually occurs?

a. Fungi rely on phototropism to aid in obtaining resources.

b. Humans regulate temperature solely by sweating and shivering.

c. Plants maintain circadian rhythms, such as timing when their stomata open.

d. Protists use the hormone endocrine-driven feedback loops to regulate processes.

e. Bacteria proliferate rapidly via mitosis to help guarantee survival of the colony.

9. Which process occurs immediately after an action potential?

a. Sodium ions rush into the cell

b. Calcium ions rush out of the cell

c. Hyperpolarization

d. Potassium ions rush into the cell

e. Repolarization

10. What important function are the roots of plants responsible for?
    a. Carrying out the Calvin cycle to produce usable energy
    b. Performing photosynthesis
    c. Conducting sugars downward through the leaves
    d. Supporting the plant body
    e. Absorbing water from the surrounding environment

11. What is the MAIN function of the respiratory system?
    a. To eliminate waste through the kidneys and bladder
    b. To exchange gas between the air and circulating blood
    c. To transform food and liquids into energy
    d. To excrete waste from the body
    e. To produce gametes for sexual reproduction and species propagation

12. Which of the following would occur in response to a change in water concentration?
    a. Phototropism
    b. Thermotropism
    c. Gravitropism
    d. Hydrotropism
    e. Chemotropism

Answer Explanations

1. A: This is a negative feedback loop that fine-tunes hormone production to maintain homeostasis. If calcium levels are too high, the hypothalamus detects it and stimulates a chain to produce TSH to lower the levels. If calcium levels are too low, the hypothalamus detects it and inhibits TRH production. Choice *E* is incorrect because thyroid hormone (T3 and T4), not TRH directly, stimulate metabolism and increase body temperature. Choice *D* is incorrect because TRH and calcium cooperate in the feedback loop as shown in the diagram below. Choice *C* is incorrect because TSH ultimately results in decreased calcium levels. Choice *B* is also false because calcitonin is regulated by TSH, not T_3 and T_4.

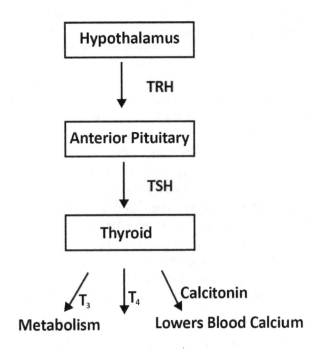

2. E: The data table is referring to rate of respiration. The more time required to use oxygen indicates a lower rate of respiration. The quicker the animal uses up oxygen, the more aerobic respiration is taking place. Ectothermic and endothermic animals are different, since endotherms have to regulate their own internal temperature, which requires more ATP because it is controlled by an elaborate feedback loop. More ATP requires higher levels of respiration. Choice *C* is attractive because it is true that smaller organisms respire faster due to their larger surface area to volume ratio. However, Choice *C* is incorrect for two reasons. First, not all lizards are smaller than the other organisms. Secondly, the dramatic nature of the difference is due to the lizards' simpler mode of temperature regulation. Choice *A* is incorrect because the small size of the lizard does not translate to a low metabolic rate, but a high one. Choice *D* is attractive because lizards bask in sunny habitats to help maintain ideal temperatures, but they lack a feedback loop that calibrates temperature, so it is incorrect. Choice *B* is incorrect because mitochondrial structure is the same among all eukaryotes — it does not change.

3. B: Photoperiodism is the phenomenon where dark exposure determines flowering, and therefore Choices *C* and *D* are wrong. Choice *B* is correct because plant B is dependent on the amount of darkness—not on the amount of light. The interrupted darkness in scenario three prevents flowering of the plant. Flowering in the winter, or when the amount of light is limited, is unusual because usually springtime is when flowers make their appearance. This could be considered an adaptation because it reduces competition for pollinators, since other flowers are not present to compete, ensuring the plant passes on its genes. Choice A is untrue because auxin is responsible for stem elongation and is not related to flowering. Choice E is not true because plant A also flowers in scenario one where it does not receive any light around midnight.

4. D: This data suggests that FSH and LH are both regulated by the same mechanism because they spike at the same time. Choice A is incorrect because while right after the FSH spike, progesterone levels increase, FSH immediately decreases and progesterone is unaffected. Choice B is incorrect because estrogen has no effect on LH in the luteal phase. Choice C is not the best choice because in the follicular phase, the levels of progesterone, LH and FSH are all low. Choice E is incorrect because during ovulation, estrogen spikes prior to LH, not the other way around.

5. A: Organisms are constantly balancing energy expenditure to do three things: maintain homeostasis, grow, and reproduce. The uterus is only necessary during pregnancy. Keeping it ready throughout a woman's entire menstrual cycle would be, using a baseball analogy, like standing at home plate in batting position when there is not a game going on. Maintaining the uterine lining when it isn't necessary is a waste of energy that could be used in other cellular processes. Choice B is incorrect because conserving energy for an upcoming event is usually associated with seasonal events such as hibernation, where energy is stored in fats before the body goes into "sleep mode." Choice C is incorrect because it is referring to reproductive animal behavior and not energy conservation and expenditure. Choice D is actually an attractive answer. If eggs were always available like sperm, a woman could theoretically have hundreds of implanted eggs, which would suck every resource and every bit of energy out of her until she died, and then the embryos couldn't live because they would starve. So, it is beneficial to release the egg monthly. The nature of the question, however, makes this choice incorrect because it is specifically asking about the hormonal regulation. To have all eggs out there at once would make the cycle hormone-independent. Choice E is nonsensical because menstruation is an increase in energy expenditure.

6. C: During vaccination, the immune response is initiated. This is called the primary, or first, immune response. Any subsequent response is called the secondary immune response because the "blueprints" to make the antibodies are circulating actively in memory B cells. The memory B cells are prepared for a "second" exposure because they can recognize the antigen. Choice C is the right answer because the memory of the vaccination stimulates B cells to proliferate. Choices A and E are incorrect because memory B cells, not antibodies or immunoglobulins (which are the same thing), are circulating. The memory B cells help create antibodies. Choice B is incorrect because B cells, not T cells, produce antibodies. Choice D is incorrect because phagocytes do not present antigens until after ingestion of a pathogen.

7. B: The question is based on absorption efficiency and surface area to volume ratio. The larger the ratio, the better the cell will be at absorption because there is more surface area for absorption to occur. The first step is to calculate the surface area to volume ratios of the different groups. This can be done by likening the microvilli structure to a rectangular prism or cylinder, which will not give exact answers, but will serve as a model that will reflect the trend. The following calculations use a rectangular prism model for calculations.

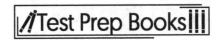

Surface area of a rectangular prism minus the side where it is attached to the small intestine is:

$$length \times width \times 4 + the\ outer\ face$$

The surface area of the outer face is equal to its width times height. The volume of the rectangular prism is:

$$length \times width \times height$$

To find the surface area to volume ratio, divide the surface area by the volume. The highest surface area to volume ratio is the most efficient microvilli, meaning it is the one absorbing the most nutrients.

Group	Surface Area $L \times W \times 4 + W \times H$	Volume $L \times W \times H$	Ratio
Unaffected	15.7	3.0	5:1
Type I diabetes	25.3	9.5	3:1
Celiac	10.4	3.2	3:1
Hypoglycemia	10.3	1.3	8:1

Choices *C* and *D* are wrong because the unaffected group does not have the largest or smallest surface area to volume ratio. Choice *A* is wrong because while the celiac group does have the smallest surface area to volume ratio, that makes it the least efficient, not the most. Choice *E* is wrong because the celiac smallest, not largest, surface area to volume ratio. Choice *B* is the best choice because a large surface area to volume ratio results in more efficient transport.

8. C: Fungi are not photosynthetic, making Choice *A* incorrect. Choice *B* is incorrect because temperature regulation is more complicated than that. For example, induction of fever as part of the immune system's response to pathogens is under the umbrella of temperature regulation. Choice *D* is incorrect because protists are single-celled organisms for the most part, and the multicellular ones are not differentiated enough to have organs; therefore, they do not have an endocrine system. Choice *E* is incorrect because bacteria do not divide mitotically; they mainly replicate asexually via binary fission.

9. A: Sodium ions rush into the cell. Choices *C* and *E* are incorrect because sodium influx causes depolarization, not hyperpolarization nor repolarization. Choices *D* and *E* are incorrect because potassium ions actually rush out of the cell (repolarization). Choice *B* is incorrect because calcium ions are not involved (until later at the synaptic knob).

10. E: Roots are responsible for absorbing water and nutrients that will get transported up through the plant. They also anchor the plant to the ground. Photosynthesis (and the Calvin cycle, which is part of photosynthesis) occurs in leaves, stems transport materials through the plant and support the plant body, and phloem moves sugars downward to the leaves.

11. B: The respiratory system mediates the exchange of gas between the air and the circulating blood, mainly by the act of breathing. It filters, warms, and humidifies the air that gets breathed in and then passes it into the blood stream. The digestive system transforms food and liquids into energy and helps excrete waste from the body. Eliminating waste via the kidneys and bladder is a function of the urinary system. The reproductive system produces gametes (sex cells) for sexual reproduction.

12. D: Tropism is a response to stimuli that causes the plant to grow toward or away from the stimuli. Hydrotropism is a response to a change in water concentration. Phototropism is a reaction to light that causes plants to grow toward the source of the light. Thermotropism is a response to changes in temperature. Gravitropism is a response to gravity that causes roots to follow the pull of gravity and grow downward, but also causes plant shoots to act against gravity and grow upward. Chemotropism is chemical-induced growth or movement.

Evolution and Diversity

Origin of Life

Abiogenesis is the term used to refer to the theoretical process by which life developed from non-living matter, such as organic compounds. Many scientists estimate that the Earth is approximately 4.5 billion years old and believe that the first living organisms developed between 3.8 and 4.1 billion years ago. There are several theories about how life formed on Earth.

One theory on the origin of life involves the creation of organic compounds from a combination of minerals from the sea and ideal atmospheric conditions. In the deep sea, hydrothermal vents release minerals from the Earth's interior along with hot water. Water from the alkaline vents on the sea floor has a high pH and provides a stable environment for organic compounds. In addition, theorists propose that the Earth's atmosphere had reducing qualities that could produce organic compounds from simpler molecules. They further believe that the warm atmosphere above volcanoes was ideal for this synthesis of organic compounds. The **primordial soup theory** proposes that larger and more complex compounds were made over time from these original organic compounds, eventually forming living organisms.

Panspermia is the concept that life came to Earth from other areas of the universe. It hypothesizes that meteoroids, asteroids, and other small objects from space landed on Earth and transferred microorganisms to the Earth's surface. This theory proposes that there were seeds of life everywhere in the universe, and when these seeds were brought to Earth, the conditions were ideal for living organisms to develop and flourish.

Protocells are small, round groups of lipids hypothesized to be responsible for the origin of life. **Vesicles** are fluid-filled compartments enclosed by a membrane-like structure. They form spontaneously when lipids, such as protocells, are added to water and have important features for the creation of living organisms, such as dividing on their own to form new vesicles and absorbing material around them. When vesicles encounter lipids, the lipids form a bilayer around the vesicle, similar to a plasma membrane of a cell. According to scientists, these cells were able to encapsulate minerals and organic molecules around them. Some of the clay that covered the primordial Earth was believed to be covered in RNA, which the vesicles could encapsulate. These simple behaviors are believed to have given rise to more complex behaviors, such as simple cell metabolism, and began resembling true cells, now known as protocells. As the cells interacted with each other, theorist propose that larger living organisms were created.

Evolutionists propose that life on Earth began with RNA. Although RNA is a genetic material, it can also be an enzyme-like catalyst, known as a **ribozyme**. Ribozymes can catalyze chemical reactions and self-replicate to make complementary copies of short pieces of RNA. According to the theory of evolution, vesicles that carried RNA could then divide and have replicated RNA in its daughter cell, increasing the amount of genetic material in the environment. These daughter cells would have been protocells. Evolutionists believe that the RNA inside of them were likely used as templates to then create more stable DNA strands. It's proposed that then from the formation of DNA, the origin of life and more complex living organisms began.

In 2015, scientists Sergei Maslov and Alexei Tkachenko expanded on this theory. They believe that the self-replicating model was cyclical and went through different phases during the day and night. During the day, the polymers would float freely, while at night, the polymer chains would join together to form longer polymers using a template, a process called **template-assisted ligation**. They believe that although the chains could join without a template, the use of a template is more efficient and reproducible for preserving the original sequences. According to these two scientists, these phases occurred at different times due to changes in the environment, such as with temperature, pH, and salinity. These factors then regulated whether the polymers would come together or float apart.

Scientific Evidence of Origin Theories

Over the years, many scientific experiments have attempted to prove that different theories about the origin of life are true. In 1953, the Miller-Urey chemical experiment simulated what they believe to be the atmospheric conditions of early Earth. It was believed that the atmosphere contained water, methane, ammonia, and hydrogen. Scientists Stanley Miller and Harold Urey showed that an electrical spark, like a bolt of lightning, helped catalyze the creation of complex organic compounds from simpler ones. They hypothesized that the complex molecules would then react with each other and the simple compounds to form even more compounds, such as formaldehyde, hydrogen cyanide, glycine, and sugars to produce life.

In support of self-replicating RNA, several scientists have tried to create the shortest RNA chain possible that can replicate itself. In the 1960s, Sol Spiegelman created a short RNA chain consisting of 218 bases that was able to replicate itself with an enzyme from a 4500 base bacterial RNA. In 1997, Manfred Eigen was able to further degrade a large RNA chain to only approximately 50 bases, which was the minimum length needed to bind a replication enzyme. Similarly, researchers at the J. Craig Venter Institute have used engineering techniques to try to create prokaryotic cells with as few genes as possible to figure out the minimal requirements for life. In 1995, they started with a microbe with the smallest genome known to humans with 470 genes and were able to take away one gene at a time to leave only 375 essential genes.

Evidence of Evolution

The Fossil Record

Fossils are the preserved remains of animals and organisms from the past, and they can elucidate the homology of both living and extinct species. Many scientists believe that fossils often provide evidence for evolution. They further propose that looking at the **fossil record** over time can help identify how quickly or slowly evolutionary changes occurred, and can also help match those changes to environmental changes that were happening concurrently.

Comparative Genetics

In **comparative genetics**, different organisms are compared at a genetic level to look for similarities and differences. Evolutionists look to DNA sequence, genes, gene order, and other structural features to look for evolutionary relationships and common ancestors between the organisms. Comparative genetics was used by scientists who proposed a link between the history of humans and chimpanzees.

Homology

Evolutionists propose that organisms that developed from a common ancestor often have similar characteristics that function differently. This similarity is known as **homology**. For example, humans, cats, whales, and bats all have bones arranged in the same manner from their shoulders to their digits.

However, the bones form arms in humans, forelegs in cats, flippers in whales, and wings in bats, and these forelimbs are used for lifting, walking, swimming, and flying, respectively. Evolutionists look to homology, believing that the similarity of the bone structure shows a common ancestry, but the functional differences are the product of evolution.

Homologous Structures

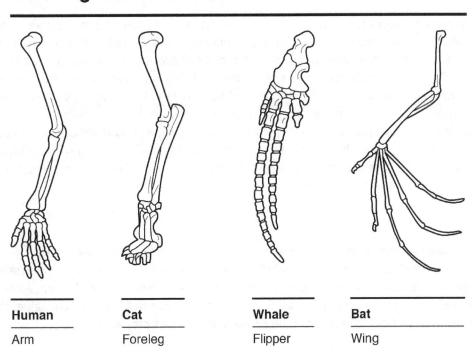

Human	Cat	Whale	Bat
Arm	Foreleg	Flipper	Wing

Patterns of Evolution

Extinction of Species

Just as speciation creates new species, the populations of some organisms shrink and eventually become extinct. A species is considered **extinct** at the death of its very last individual. Some species cannot adapt to changes in their environment and may also find themselves weaker than new species. Speciation may generate a strong species that preys on weaker species until they are extinct. Competition between species for limited resources may also wipe out a weaker species.

Reproductive Isolation Can Impact Speciation

One important distinguishing factor in the formation of two species is their reproductive isolation. Species are characterized by their members' ability to breed and produce viable offspring. When speciation occurs and new species are formed, there must be a biological barrier that prevents the two species from producing viable offspring.

Following speciation, there are two types of **reproductive barriers** that keep the two populations from mating with each other. These are classified as either prezygotic barriers or postzygotic barriers. **Prezygotic barriers** prevent fertilization via habitat isolation, temporal isolation, and behavioral isolation. Through habitat isolation, two species may inhabit the same area but don't often encounter each other. **Temporal isolation** is when species breed at different times of the day, during different seasons, or during different years, so their mating patterns never coincide. **Behavioral isolation** refers to mating rituals that prevent an organism from recognizing a different species as potential mate.

Other prezygotic barriers block fertilization after a mating attempt. **Mechanical isolation** occurs when anatomical differences prevent fertilization. **Gametic isolation** occurs when the gametes of two species are incompatible.

Postzygotic barriers contribute to reproductive isolation after fertilization. The mixed-species offspring are called **hybrids**. Once the hybrid zygote is formed, it may have reduced viability because of differing numbers of chromosomes between the parents. When meiosis takes place, normal gamete cells aren't produced in the offspring. These hybrids may not survive, or if they do, they are very frail. If a hybrid survives and becomes a healthy animal, it may have reduced fertility. For example, when a male donkey mates with a female horse, they produce a mule. Mules are healthy but sterile. When a female donkey mates with a male horse, they produce a hinny, which is also sterile. **Hybrid breakdown** is another postzygotic barrier. In some cases, hybrids that are produced are actually viable and fertile for the first generation. However, when the hybrids mate with each other or one of the parent species, the subsequent generations are weak or sterile.

Continuous Change

Although natural selection supports the survival and reproduction of the individuals best suited for their environment, there variation always remains within a population. This idea is known as balancing selection. Balancing selection maintains the variation in a population through the heterozygote advantage and frequency-dependent selection. Because the environment is always changing, the characteristics necessary to increase or maintain survival of a species may also change.

Heterozygote Advantage

The **heterozygote advantage** refers to when individuals that are heterozygous for a particular trait are more fit for their environment than individuals that are homozygous for either variation of the trait. An example is the gene that codes for the β-subunit of hemoglobin, the oxygen-carrying protein of red blood cells. The homozygous genotype causes sickle-cell disease, which distorts the shape of red blood cells so they block the flow of blood. The heterozygous genotype for this subunit of hemoglobin protects humans against the most severe effects of malaria because the body can quickly destroy the distorted red blood cells and get rid of the parasites carrying malaria. The human population is now predominantly heterozygous for this gene.

Frequency-Dependent Selection

In **frequency-dependent selection**, the occurrence of a phenotype depends on how common it is in the population. If a phenotype becomes too common, it may no longer benefit the survival and reproduction of the species. For example, there are scale-eating fish in Lake Tanganyika in Africa that can be either right-mouthed or left-mouthed. Left-mouthed fish attack their prey's right flank, and right-mouthed fish attack their prey's left flank. The prey species learn to protect themselves from the type of scale-eating fish most common in the lake. Therefore, the dominance of the left- and right-mouthed fish in the lake changes over time and keeps the species from becoming extinct due to lack of prey.

Balancing selection keeps both alleles of the gene present in the scale-eating fish population, and is propose by some to play a large role in their evolution.

Natural Selection

Biological evolution is the concept that a population's gene pool changes over generations. According to this concept, populations of organisms evolve, not individuals, and over time, genetic variation and mutations lead to such changes.

Charles Darwin developed a scientific model of evolution based on the idea of **natural selection**. When some individuals within a population have traits that are better suited to their environment than other individuals, those with the better-suited traits tend to survive longer and have more offspring. The survival and inheritance of these traits through many subsequent generations lead to a change in the population's gene pool. According to natural selection, traits that are more advantageous for survival and reproduction in an environment become more common in subsequent generations.

Evolutionary Fitness

Sexual selection is a type of natural selection in which individuals with certain traits are more likely to find a mate than individuals without those traits. This can occur through direct competition of one sex for a mate of the opposite sex. For example, larger males may prevent smaller males from mating by using their size advantage to keep them away from the females. Sexual selection can also occur through mate choice. This can happen when individuals of one sex are choosy about their mate of the opposite sex, often judging their potential mate based on appearance or behavior. For example, female peacocks often mate with the showiest male with large, beautiful feathers. In both types of sexual selection, individuals with some traits have better reproductive success, and the genes for those traits become more prevalent in subsequent populations.

Adaptations are Favored by Natural Selection

The theory of **adaptation** is defined as an alteration in a species that causes it to become more well-suited to its environment. Charles Darwin's idea of natural selection explains *how* populations change—adaption explains *why*. It increases the probability of survival, thus increasing the rate of successful reproduction. As a result, an adaptation becomes more common within the population of that species.

For examples, bats use reflected sound waves (echolocation) to prey on insects, and chameleons change colors to blend in with their surroundings to evade detection by its prey and predators.

Adaptive radiation refers to rapid diversification within a species into an array of unique forms. It may occur as a result of changes in a habitat creating new challenges, ecological niches, or natural resources.

Darwin's finches are often thought of as an example of the theory of adaptive radiation. Charles Darwin documented 13 varieties of finches on the Galapagos Islands. Each island in the chain presented a unique and changing environment, which was believed to cause rapid adaptive radiation among the finches. There was also diversity among finches inhabiting the same island. Darwin believed that as a result of natural selection, each variety of finch developed adaptations to fit into its native environment.

A major difference in Darwin's finches had to do with the size and shapes of beaks. The variation in beaks allowed the finches to access different foods and natural resources, which decreased competition

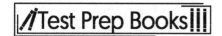

and preserved resources. As a result, various finches of the same species were allowed to coexist, thrive, and diversify.

Finches had:

- Short beaks, which were suited for foraging for seeds
- Thin, sharp beaks, which were suited for preying on insects
- Long beaks, which were suited for probing for food inside plants

Darwin believed that the finches on the Galapagos Islands resulted from chance mutations in genes transmitted from generation to generation.

Populations in Hardy-Weinberg Equilibrium

All populations have genetic diversity, but according to evolutionists, some populations aren't evolving. The **gene pool** consists of all copies of every allele at every locus in every member of a population. If the allele and genotype frequencies of a population don't change between generations, the population is in a **Hardy-Weinberg (HW) equilibrium**, named for the British mathematician and German physician who came up with the concept in 1908. There are five conditions that must be met for a HW equilibrium: (1) a large population size, (2) absence of migration, (3) no net mutations, (4) random mating, and (5) absence of selection.

The HW equation calculates the frequency of phenotypes in a population that isn't evolving and is written as follows: $p^2 + 2pq + q^2 = 1$, where p is the frequency of one allele, q is the frequency of the other allele, and pq is the frequency of the alleles mixing. The sum of p and q must be equal to 1. As in the figure below, in a given population of wildflowers, the frequency of the red flower allele (p) is 80%, and the frequency of the white flower allele (q) is 20%. Therefore, $p = 0.8$ and $q = 0.2$. In a non-evolving population, the frequency of red flowers would be $p^2 = 0.64 = 64\%$, the frequency of pink flowers as a mix of red flower and white flower alleles would be $2pq = 0.32 = 32\%$, and the frequency of white flowers would be $q^2 = 0.04 = 4\%$. If the frequency of any flower color doesn't match the calculations from the HW equation, then according to this equation, the population is evolving.

Parameters for Natural Selection

There are three important points to remember about natural selection. Although natural selection occurs due to an individual organism's relationship to its environment, it is a population—not individuals—that change over time. Second, natural selection only affects heritable traits that vary within a population. If all individuals within a population share an identical trait, natural selection cannot occur, and that trait will not be modified. Lastly, favored traits are always changing. The environment is an important factor in natural selection, so if the environment changes, a trait that was previously favored may no longer be beneficial. Natural selection is a fluid process that is always at work.

Environmental Changes Serve as Selective Mechanisms

The environment constantly changes, which drives selection. Although an individual's traits are determined by their **genotype**, or makeup of genes, natural selection more directly influences **phenotype**, or observable characteristics. The outward appearance or ability of individuals affects their ability to adapt to their environment and survive and reproduce. Phenotypic changes occurring in a population over time are accompanied by changes in the gene pool.

The classic example of this is the peppered moth. It was once a light-colored moth with black spots, though a few members of the species had a genetic variation resulting in a dark color. When the Industrial Revolution hit London, the air became filled with soot and turned the white trees darker in color. Birds were then able to spot and eat the light-colored moths more easily. Within just a few months, the moths with genes for darker color were better able to avoid predation. Subsequent generations had far more dark-colored moths than light ones. Once the Industrial Revolution ended and the air cleared, light-colored moths were better able to survive, and their numbers increased.

Causes of Phenotypic Variations

There are three ways in which phenotypes change over time due to natural selection: directional selection, disruptive selection, and stabilizing selection. **Directional selection** occurs when an extreme phenotypic variation is favored. This generally happens when a population's environment changes or the population migrates to a new habitat. When the Galapagos Islands suffered a drought, finches with larger beaks were able to eat the larger, tougher seeds that became abundant. Thus, finches with that phenotype survived and reproduced more often, and that trait became more prevalent in subsequent generations. **Disruptive selection** occurs when both extremes of a phenotype are favored. Finches in Cameroon have either large beaks or small beaks. The large-beaked birds are efficient at eating large, tough seeds; the small-beaked birds are adept at eating small seeds. Birds with medium-sized beaks were not adept at eating either size of seed, so selection favored the other finches. **Stabilizing selection** occurs when neither extreme phenotype is favored, and the intermediate phenotype is best suited for adapting to the population's environment. If mice live in an environment with a mix of light and dark colored rocks, mice with an intermediate fur color are favored. Neither light nor dark fur will be selected.

The Effect of Phenotypic Variations on Fitness

Geneticists have a specialized definition of fitness. They use the term to denote an organism's capacity to survive, mate, and reproduce. This ultimately equates to the probability or likelihood that the organism will be able to pass on its genetic information to the next generation. Fitness does not mean the strongest, biggest, or most dangerous individual. A subtler combination of anatomy, physiology, biochemistry, and behavior determine genetic fitness.

Another way to understand genetic fitness is by knowing that phenotypes affect survival and the ability to successfully reproduce. Phenotypes are genetically determined and genes contributing to fitness tend to increase over time.

Thus, the "fittest" organisms survive and pass on their genetic makeup to the next generation. This is what Darwin meant by "survival of the fittest," which is the cornerstone of Darwin's theory of evolution.

Influence of Phenotype on Genotype

Natural selection provides processes that tend to increase a population's adaptive abilities. The strongest phenotypes survive, prosper, and pass on their genetic code to the next generation. The new generation of phenotypes has a fitter genotype because they have inherited more adaptive characteristics.

Thus, the fit thrive while the weak become extinct over time. This pattern, when repeated over many generations, develops strong, fit phenotypes that survive and reproduce offspring who are as fit as or fitter than their parents. Sometimes this apparently inexorable movement can be modified by drastic external conditions such as wide variations in climate. It is important to remember that all extinct species were once fit and adapted to their environments. Unforeseen circumstances always have the potential to cause chaos in the physical world.

Other Causes of Genetic Changes in Species and Populations

While the concept of natural selection focuses on changes over time in relation to the environment, there are other circumstances when change occurs randomly. Not all genetic changes relate to survival and reproduction. **Genetic drift** is the idea that the alleles of a gene can change unpredictably between generations due to chance events. If certain alleles are lost between generations, the genetic diversity of the population decreases because that genetic variation is lost forever. For example, a population of wildflowers consists of red flowers (*RR* and *Rr*) and white flowers (*rr*). If a large animal destroys all of the white wildflowers, the subsequent generation could be left with no (or far fewer) alleles for white flowers. Genetic drift has the greatest effect on smaller populations. Certain alleles can be over- or under-represented, even if they are not advantageous. In addition, harmful alleles can become fixed if their normal counterpart becomes extinct.

A **population bottleneck** is a type of genetic drift. This occurs when a population significantly decreases, usually due to a sudden change in the environment such as a flood or a fire. In the surviving population, certain alleles may be over- or under-represented, and others may be completely missing. Even if the surviving population returns to its original size, it will lack the genetic diversity of the original population. The **founder effect** is a special case of the bottleneck effect. It occurs when a few individuals become separated from the larger population and form their own new population. The frequency of non-dominant alleles may increase in the new smaller population, as may the frequency of inherited disorders due to a lack of dominant alleles that would keep the disorder inactive.

Speciation

Speciation is the process by which one species splits into two or more species. There are two main types of speciation that can occur: allopatric and sympatric. **Allopatric speciation** happens because of geographic separation. One population is divided into two subpopulations. If a large lake becomes divided into two smaller lakes during a drought, each lake has its own population of a species of fish that cannot intermingle with the fish in the other lake. This type of speciation can also occur when a subgroup of a population migrates to a new geographic area. When the genes of these two subpopulations are no longer mixing, new mutations can arise, and natural selection and genetic drift can take place.

In **sympatric speciation**, two or more species arise without a geographic barrier. Instead, gene flow among the population is reduced by polyploidy, sexual selection, or habitat differentiation. **Polyploidy** results when cell division during reproduction creates an extra set of chromosomes; it's more common in plants than animals. In sexual selection, organisms of one sex choose their mate based on certain traits. If there's high selection for two extreme variations of a trait, sympatric speciation may occur. For example, the females of two species of cichlid fish choose their mates based on the color of their backs—one has a blue-tinged back and the other has a red-tinged back. **Habitat differentiation** occurs when a subpopulation exploits a resource that's not used by the parent population. The North American maggot fly originally used the hawthorn tree as its habitat. A subsequent generation of this fly chose to colonize apple trees instead, creating two populations.

Both allopatric and sympatric speciation can occur quickly or slowly, and may involve a few or many genetic changes between the species.

Classification and Diversity of Organisms

Taxonomy is the science behind the biological names of organisms. Biologists often refer to organisms by their Latin scientific names to avoid confusion with common names, such as with fish. Jellyfish, crayfish, and silverfish all have the word "fish" in their name, but belong to three different species. In the eighteenth century, Carl Linnaeus invented a naming system for species that included using the Latin scientific name of a species, called the **binomial**, which has two parts: the **genus**, which comes first, and the **specific epithet**, which comes second. Similar species are grouped into the same genus. The Linnaean system is the commonly used taxonomic system today and, moving from comprehensive similarities to more general similarities, classifies organisms into their species, genus, family, order, class, phylum, kingdom, and domain. *Homo sapiens* is the Latin scientific name for humans.

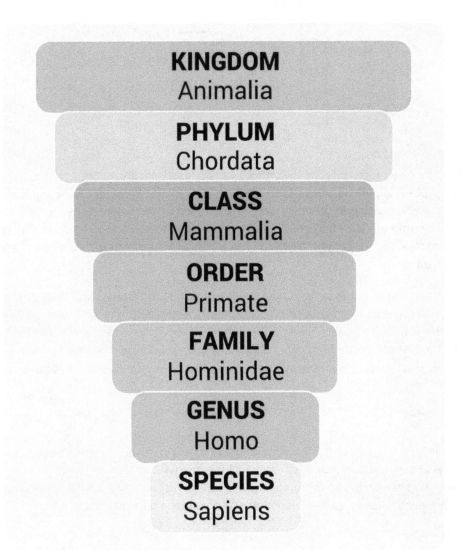

Phylogenetic trees are branching diagrams that are used to represent the believed evolutionary history of a species. The branch points most often match the classification groups set forth by the Linnaean system. Using this system helps elucidate the relationship between different groups of organisms. The diagram below is that of an empty phylogenetic tree:

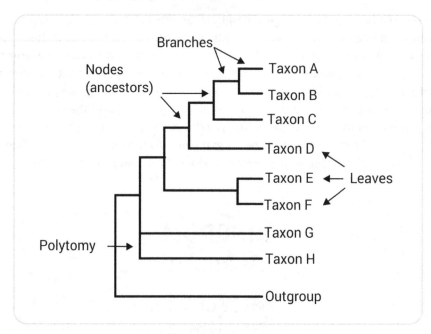

Each branch of the tree represents the divergence of two species from a common ancestor. For example, the coyote is known as *Canis latrans* and the gray wolf is known as *Canis lupus*. They share the same genus but evolved into two different species and could be represented by Taxon A and Taxon B extending from a common *Canis* branch. Groups or organisms that share a common ancestor or node in the diagram are known as **sister taxa**. A branch point leading to more than two descendent groups is known as a **polytomy**.

There are three important points to remember about phylogenetic trees. First, their purpose is to show a pattern of descent, not to indicate phenotypic similarity between organisms. Second, the diagram doesn't represent time. Two taxons from the same ancestor may have become their own species at different points in history and at different rates. Lastly, taxons that are next to each other didn't come from each other. The tree diagram only indicates that they came from the same ancestor, which is noted at their shared node.

Cladograms

Cladistics is a method of classifying organisms based primarily on their common ancestry. Using this method, species are grouped into **clades**, which include one ancestral species and all of its descendants. Some assume that species with similar traits are related. However, these similarities may appear by analogy, which means that the species were subject to similar a natural selection process but don't

share a common ancestor. Cladograms help discern the difference between analogous features and homologous features. Below is an example of a cladogram:

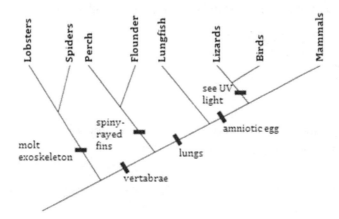

While a phylogenetic tree diagrams an organism's believed evolutionary history, a cladogram specifies the characteristics that change within the descendent groups, making it easy to see the homology of traits between related species.

Characteristics of Bacteria, Animals, Plants, Fungi, and Protists

As discussed earlier, there are two distinct types of cells that make up most living organisms: prokaryotic and eukaryotic. Bacteria (and archaea) are classified as prokaryotic cells, whereas animal, plant, fungi, and protist cells are classified as eukaryotic cells.

Although animal cells and plant cells are both eukaryotic, they each have several distinguishing characteristics. **Animal cells** are surrounded by a plasma membrane, while **plant cells** have a cell wall made up of cellulose that provides more structure and an extra layer of protection for the cell. Animals use oxygen to breathe and give off carbon dioxide, while plants do the opposite—they take in carbon dioxide and give off oxygen. Plants also use light as a source of energy. Animals have highly developed sensory and nervous systems and the ability to move freely, while plants lack both abilities. Animals, however, cannot make their own food and must rely on their environment to provide sufficient nutrition, whereas plants do make their own food.

Fungal cells are typical eukaryotes, containing both a nucleus and membrane-bound organelles. They have a cell wall, similar to plant cells; however, they use oxygen as a source of energy and cannot perform photosynthesis. They also depend on outside sources for nutrition and cannot produce their own food. Of note, their cell walls contain **chitin**.

Protists are a group of diverse eukaryotic cells that are often grouped together because they do not fit into the categories of animal, plant, or fungal cells. They can be categorized into three broad categories: protozoa, protophyta, and molds. These three broad categories are essentially "animal-like," "plant-like," and "fungus-like," respectively. All of them are unicellular and do not form tissues. Besides this simple similarity, protists are a diverse group of organisms with different characteristics, life cycles, and cellular structures.

Practice Questions

1. Charles Darwin's theory of evolution is based on what type of selection?
    a. Natural selection
    b. Sexual selection
    c. Disruptive selection
    d. Stabilizing selection
    e. Allopatric selection

2. Which of the following phrases best fits into the following sentence?
Natural selection makes individuals _____ their environments.
    a. less fit to
    b. less affected by
    c. grow faster in
    d. eat more in
    e. more adaptable to

3. Which type of selection describes the finches of the Galapagos Islands developing larger beaks so they are able to eat the larger, tougher seeds that became abundant after a drought?
    a. Sexual
    b. Stabilizing
    c. Directional
    d. Disruptive
    e. Adaptable

4. When mice develop an intermediate fur color instead of light or dark fur, what type of selection is occurring?
    a. Disruptive
    b. Stabilizing
    c. Directional
    d. Sexual
    e. Adaptable

5. The founder effect occurs when which of the following occur?
    a. A new species suddenly fills an open niche.
    b. A new species is developed.
    c. Individuals develop an extreme phenotype through natural selection.
    d. There is a sudden change in environmental conditions.
    e. A few individuals become separated from the larger population and form a new population.

6. Which circumstance of random change leaves a population less diverse than its original composition?
    a. A geographical barrier
    b. Founder effect
    c. A strong wind
    d. Bottleneck effect
    e. Speciation

7. What do evolutionary theorists believe the Hardy-Weinberg equation tells us?
   a. Whether a population is evolving
   b. The type of natural selection occurring
   c. If genetic drift is altering a population
   d. The size of the population
   e. The expected rate of population growth.

8. What is an adaptation?
   a. The original traits found in a common ancestor
   b. Changes that occur in the environment
   c. When one species begins behaving like another species
   d. An inherited characteristic that enhances survival and reproduction
   e. Changes that occur in an individual during the aging process

9. What is the broadest, or LEAST specialized, classification of the Linnaean taxonomic system?
   a. Species
   b. Family
   c. Domain
   d. Phylum
   e. Kingdom

10. According to evolutionary theory, what are vestigial structures?
   a. Anatomical structures that stick out of the body
   b. Structures found only in the foot
   c. Structures found only in the hand
   d. Anatomical structures that are still present but no longer have a function
   e. Structures that give a species a phenotypical advantage in their environment

11. Which taxonomic system is commonly used to describe the hierarchy of similar organisms today?
   a. Aristotle system
   b. Linnaean system
   c. Cesalpino system
   d. Darwin system
   e. The Hardy-Weinberg system

12. What is the Latin specific name for humans?
   a. *Homo sapiens*
   b. *Homo erectus*
   c. *Felis catus*
   d. *Canis familiaris*
   e. *Homo habilis*

13. What do phylogenetic trees tell us about a species?
   a. The genetic contribution of each allele that an offspring inherits
   b. The believed size of the population
   c. How many alleles exist for a specific trait of the species
   d. Their eye color
   e. Their proposed evolutionary history

14. How do cladograms differ from phylogenetic trees?
    a. They include pictures of the species
    b. They specify common species names instead of Latin names
    c. They are not branching diagrams
    d. They specify the characteristics that changed in descendent groups, creating new species
    e. They show the specific alleles that each parent had, and the potential crosses that could result

15. What is speciation?
    a. The process by which a species becomes extinct
    b. The relationship between two species as described by a phylogenetic tree
    c. The act of drawing a phylogenetic tree
    d. The process by which a predator overtakes its prey
    e. The process by which one species splits into two or more species

16. In allopatric speciation, what causes a species to split in two?
    a. A geographic barrier
    b. Natural selection
    c. A change in environment
    d. Interspecies mating
    e. Competition

17. Sexual selection stops gene flow and causes what type of speciation?
    a. Allopatric
    b. Natural
    c. Sympatric
    d. Sexual
    e. Prezygotic

18. In addition to pollution and deforestation, what other human activities can cause extinction?
    a. Urbanization and overfishing
    b. Urbanization and competition
    c. Predation and urbanization
    d. Overfishing and competition
    e. Predation and competition

19. When is a species considered extinct?
    a. At the death of the last individual
    b. When there are fewer than 10 individuals remaining
    c. When the species has not had any individuals alive for 5 years
    d. When exactly 2 individuals remain
    e. When the death rate exceeds the birth rate for at least 5 years

20. Which is an example of a prezygotic reproductive barrier?
    a. Changing environments
    b. Three species inhabiting the same area
    c. Large feathers
    d. Habitat isolation
    e. Temperature changes

21. What type of barrier leads to reproductive isolation after two species mate and produce a hybrid offspring?
   a. Postzygotic barrier
   b. Habitat isolation
   c. Prezygotic barrier
   d. Behavioral isolation
   e. Mechanical isolation

22. Which concept maintains the variation in a population even as natural selection occurs?
   a. Genetic drift
   b. Sexual selection
   c. Postzygotic barriers
   d. Balancing selection
   e. Bottleneck

23. Which idea about the origin of life hypothesizes that microorganisms were transferred to Earth from other objects in the solar system?
   a. Primordial soup
   b. Self-replicating RNA
   c. Endosymbiosis
   d. Protocells
   e. Panspermia

24. Protocells are an essential vesicle for replicating what material essential to a hypothesis about the origin of life on Earth?
   a. Clay
   b. RNA
   c. Chromosomes
   d. Carbohydrates
   e. Chloroplasts

25. What did the Miller-Urey experiment simulate to test the creation of complex compounds from simpler ones that are believed to have existed on the surface of early Earth?
   a. The atmospheric conditions
   b. The Earth's interior temperature
   c. The oceanic environment
   d. The number of leaves on each tree
   e. The nature of algae in deep sea trenches

26. What is a driving force behind why speciation can occur?
   a. Geographic separation
   b. Seasons
   c. Daylight
   d. A virus
   e. Warm temperatures

Answer Explanations

1. A: Charles Darwin founded the theory of natural selection. He believed that stronger individuals would continue to thrive while weaker individuals would die off.

2. E: Natural selection is the idea that individuals within a population can survive longer and with higher reproduction rates based on specific traits that they've inherited that make them better matched to their environment.

3. C: Natural selection acts on the phenotypes of individuals. Directional selection occurs when one extreme of the phenotypic variations is favored. This generally happens when a population's environment changes or the population migrates to a new habitat.

4. B: Stabilizing selection occurs when neither extreme phenotype is favored and the intermediate phenotype is most suitable for adapting to the population's environment. If mice live in an environment with a mix of light- and dark-colored rocks, their fur will be an intermediate color. Neither light nor dark fur will be selected.

5. E: The founder effect occurs when a few individuals from a population become separated from the larger population and form their own new population. This may occur when a storm blows a few individuals to a new island or habitat. The frequency of non-dominant alleles may increase in the new smaller population, as well as the frequency of inherited disorders due to a lack of dominant alleles that would keep the disorder inactive.

6. D: The bottleneck effect occurs when there's a sudden change in the environment. A flood or a fire could drastically decrease the size of a population. In the surviving population, certain alleles may be over- or under-represented, and others may be completely gone. Even if the surviving population reaches back to its original size over time, it will lack the genetic diversity of the original population.

7. A: All populations have genetic diversity, but according to evolutionists, that doesn't guarantee the population is evolving. In order to assess whether a population is evolving, scientists use a mathematical equation to calculate the phenotypes of a non-evolving population. The results of that equation can be compared to the actual phenotypes seen in the population. If the allele and genotype frequencies of a population aren't changing between generations, the population is described as being in a Hardy-Weinberg (HW) equilibrium.

8. D: Charles Darwin based the idea of adaptation around his original concept of natural selection. He believed that evolution occurred based on three observations: the unity of life, the diversity of life, and the suitability of organisms to their environments. There was unity in life based on the idea that all organisms descended from a common ancestor. Then, as the descendants of common ancestors faced changes in their environments or moved to new environments, they began adapting new features to help them. This concept explained the diversity of life and how organisms were matched to their environments. Natural selection helps to improve the fit between organisms and their environments by increasing the frequency of features that enhance survival and reproduction.

9. C: In the Linnaean system, organisms are classified as follows, moving from comprehensive and specific similarities to fewer and more general similarities: domain, kingdom, phylum, class, order, family, genus, and species. A popular mnemonic device to remember the Linnaean system is "Dear King Philip came over for good soup."

10. D: Evolutionists propose that structures present in descendent species that no longer have a function are known as vestigial structures. For example, some snakes still have pelvis and leg bones that descended from ancestors that walked.

11. B: The Linnaean system is the commonly used taxonomic system today. It classifies species based on their similarities and moves from comprehensive to more general similarities. The system is based on the following order: species, genus, family, order, class, phylum, and kingdom.

12. A: *Homo* is the human genus. *Homo sapiens* is the only remaining species in the *Homo* genus.

13. A: Phylogenetic trees are used to illustrate the believed evolutionary history of a species. They are branching diagrams, and the branch points most often match the classification groups set forth by the Linnaean system. Using this system helps elucidate the relationship between different groups of organisms.

14. D: Cladograms help discern the difference between analogous features and homologous features. While a phylogenetic tree diagrams the believed evolutionary history of an organism, a cladogram specifies the characteristics that change within descendent groups, making it is easy to see the homology of certain traits between related species.

15. E: Speciation is the process by which one species splits into two or more species.

16. A: Allopatric speciation happens because of geographic separation. One population is divided into two subpopulations. If a drought occurs and a large lake divides into two smaller lakes, each lake is left with its own population that cannot intermingle with the population of the other lake. This type of speciation can also occur when a subgroup of a population migrates to a new geographic area.

17. C: In sympatric speciation, two or more species arise without a geographic barrier. In sexual selection, organisms choose their mate based on certain traits. If there is high selection for two extreme variations of a trait, sympatric speciation may occur. The females of two distinct species of cichlid fish choose their male mates based on the color of their backs—one has a blue-tinged back and the other has a red-tinged back.

18. A: Humans are responsible for many harmful changes to the environment. Deforestation and urbanization cause species to lose their adapted environment all together. Overfishing and overhunting greatly decrease the number of individuals in a species. Pollution also cause destructive changes to the environment to which some species may be unable to adapt.

19. A: A species is considered extinct at the death of its very last individual. If any number of individuals are still alive, the species may be considered endangered, but not extinct.

20. D: Prezygotic barriers prevent fertilization. These include habitat isolation, temporal isolation, and behavioral isolation. Two species may inhabit the same area but don't encounter often each other, which is habitat isolation.

21. A: Postzygotic barriers contribute to reproductive isolation after fertilization. The mixed species offspring are called hybrids. Once the hybrid zygote is formed, it may have reduced viability. These hybrids may not survive, and if they do, they are very frail. If a hybrid survives, it may have reduced fertility. In some cases, hybrids are actually viable and fertile for the first generation. However, when the hybrids mate with each other or with one of the parent species, subsequent generations are weak or sterile.

22. D: Although the idea of natural selection supports the survival and reproduction of the individuals best suited for their environment, there's always variation within a population. This idea is known as balancing selection, which maintains the variation in a population through the heterozygote advantage and frequency-dependent selection. As the environment is always changing, the characteristics necessary to increase or maintain survival and reproduction of a species may also change.

23. E: Panspermia is the idea that life came to Earth from other areas of the universe. It hypothesizes that meteoroids, asteroids, and other small objects from the solar system landed on Earth and transferred microorganisms to the Earth's surface. This theory proposes that there were seeds of life everywhere in the universe, and when these seeds were brought to Earth, the conditions were ideal for living organisms to develop and flourish.

24. B: Protocells are small, round groups of lipids that are hypothesized to be responsible for the origin of life. They are proposed to have been formed by vesicles, which are fluid-filled compartments enclosed by a membrane-like structure. Vesicles form spontaneously when lipids, such as protocells, are added to water. Some of the clay that covered the Earth is believed to contain RNA, which the vesicles could encapsulate. Vesicles that carried RNA could then divide and have replicated RNA in its daughter cell, increasing the amount of genetic material in the environment. The RNA inside of them may have been used as templates to create more stable DNA strands. Then, according to evolutionary theory, after the formation of DNA, the origin of life and more complex living organisms began.

25. A: In 1953, the Miller-Urey experiment attempted to simulate the atmospheric conditions of early Earth. It was believed that the atmosphere contained water, methane, ammonia, and hydrogen. Scientists Stanley Miller and Harold Urey believed an electrical spark, such as a bolt of lightning, helped catalyze the creation of complex organic compounds from simpler ones. The complex molecules would then react with each other and the simple compounds to form even more compounds, such as formaldehyde, hydrogen cyanide, glycine, and sugars.

26. A: Speciation is the method by which one species splits into two or more species. In allopatric speciation, one population is divided into two subpopulations. If a drought occurs and a large lake becomes divided into two smaller lakes, each lake is left with its own population that cannot intermingle with the population of the other lake. When the genes of these two subpopulations are no longer mixing with each other, new mutations can arise and natural selection can take place.

Dear SAT Biology E/M Test Taker,

We would like to start by thanking you for purchasing this study guide for your SAT Biology E/M exam. We hope that we exceeded your expectations.

Our goal in creating this study guide was to cover all of the topics that you will see on the test. We also strove to make our practice questions as similar as possible to what you will encounter on test day. With that being said, if you found something that you feel was not up to your standards, please send us an email and let us know.

We would also like to let you know about other books in our catalog that may interest you.

SAT Math 1

This can be found on Amazon: amazon.com/dp/1628454717

SAT Literature

amazon.com/dp/1628456280

SAT

amazon.com/dp/1628458984

ACCUPLACER

amazon.com/dp/162845945X

AP Biology

amazon.com/dp/1628456221

We have study guides in a wide variety of fields. If the one you are looking for isn't listed above, then try searching for it on Amazon or send us an email.

Thanks Again and Happy Testing!
Product Development Team
info@studyguideteam.com

FREE Test Taking Tips DVD Offer

To help us better serve you, we have developed a Test Taking Tips DVD that we would like to give you for FREE. **This DVD covers world-class test taking tips that you can use to be even more successful when you are taking your test.**

All that we ask is that you email us your feedback about your study guide. Please let us know what you thought about it – whether that is good, bad or indifferent.

To get your **FREE Test Taking Tips DVD**, email freedvd@studyguideteam.com with "FREE DVD" in the subject line and the following information in the body of the email:

    a. The title of your study guide.

    b. Your product rating on a scale of 1-5, with 5 being the highest rating.

    c. Your feedback about the study guide. What did you think of it?

    d. Your full name and shipping address to send your free DVD.

If you have any questions or concerns, please don't hesitate to contact us at freedvd@studyguideteam.com.

Thanks again!

CPSIA information can be obtained
at www.ICGtesting.com
Printed in the USA
BVHW011145150121
597948BV00012B/118